Illustration by Camille Rose Garcia

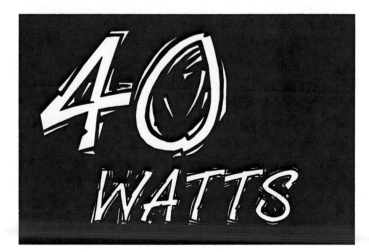

FROM
NOWHERE

A Journey into Pirate Radio

SUE CARPENTER

SCRIBNER

New York   London   Toronto   Sydney

SCRIBNER
1230 Avenue of the Americas
New York, NY 10020

Copyright © 2004 by Sue Carpenter

SCRIBNER and design are trademarks of
Macmillan Library Reference USA, Inc., used under license
by Simon & Schuster, the publisher of this work.

For information about special discounts for bulk purchases,
please contact Simon & Schuster Special Sales:
1-800-456-6798 or business@simonandschuster.com

Designed by Lauren Simonetti
Text set in Minion

Manufactured in the United States of America

1   3   5   7   9   10   8   6   4   2

Library of Congress Cataloging-in-Publication Data

Carpenter, Sue, 1966–
40 watts from nowhere : a journey into pirate radio / Sue Carpenter
p. cm.
1. Pirate radio broadcasting—California—History. I. Title: Forty watts
from nowhere. II. Title.
HE8697.65.C37 2004
384.54—dc22
2003059193

ISBN: 978-1-4165-6960-X

For Chris and Carpenter

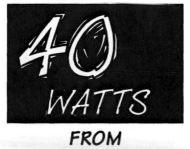

FROM

NOWHERE

INTRODUCTION

It's a Friday, about 11 A.M., when the station turns to static. My pulse doubles in an instant. I can't believe they did it *again*. I don't even know who "they" are, but it's the second time they've turned off the transmitter in two days.

Yesterday, they figured out which antenna and transmitter were ours from the half-dozen others scattered on top of the same high-rise, but they were gone before we could get there and identify them. Today, I'm not going to let them get away. I run out the door and kick start my beat-up Honda. Ordinarily I'd let the little junker warm up, but there's no time. I keep the choke on and speed down the street, revving the throttle hard.

I'm about to shift into second when a red light stops me. I train my eye on the signal and peel out the instant it goes green, taking the corner so fast my left knee almost scrapes the ground. My front tire is inches from rear-ending a Mercedes. I can't believe this traffic. It isn't rush hour. It isn't even lunchtime. Where did all these cars come from? And why are they moving so slowly? I scream at them through my face shield. They need to get out of my way.

Don't they understand what's happening here? KBLT, the radio station I built from scratch, the station I've sacrificed my apartment for—and my sanity—might be permanently kaput. I lay on the horn. I can't stand these L.A. drivers and their lane-hogging SUVs. I can't squeeze through. I pass three cars in the turning lane, praying the cops won't bust me for reckless driving.

Three glorious years—well, sometimes glorious years—of squatting on the FM dial. It can't be over. What have we done wrong? Who turned us in? Why now?

This Friday morning is no different from any other at my radio station: Eddie knocks on the door, I let him in, and he spins dub, krautrock, or whatever other music he wants for whoever's tuned in. Two hours later Hassan stops by with his crate of classic jazz, and so it goes, nonstop, around the clock, twenty-four hours each day. So what if I don't have a license to operate? I just couldn't scrape together the $100 million I needed to buy my way onto the FM band in L.A. It can't be so wrong to co-opt a little underutilized air space so music lovers can show off their record collections. It makes no sense that that's illegal.

I focus on the tower at Sunset and Vine two miles away. A jumble of antennas and satellite dishes clutter its roof. Shit. There's something moving up there. There are *people* moving. I feel sick. I need to calm down. If I don't pull it together I'm going to slip under the fender of a Mack truck and sever a leg.

Okay. One mile. Just one more mile and I'm there. Jesus, the sun is bright. In my rush, I didn't manage to grab my sunglasses. I'm squinting and can barely make them out in the distance, but two tiny ant people are wriggling somewhere in the vicinity of the KBLT antenna. This just might be the end.

CHAPTER 1

"Good morning, law offices." My voice is friendly, calm, professional—as it should be. When you repeat the same thing eight thousand times, you work out the kinks.

"No, he isn't in right now," I say, pen hovering over a hot-pink message pad. I'm talking about my boss, Allan Schwartz, an attorney who makes his mortgage defending real estate insurance companies from mostly frivolous lawsuits. Today he's in court defending his client from an angry home buyer who claims he wasn't informed about the railroad tracks running along the back edge of his property.

"Yes, I understand. Would you like to leave a message?" I take down the caller's name, number, and the time, then rip the sheet from its pad and add it to a growing stack.

All right. Where was I? Ah, yes, typing. I pick up my earphones and continue transcribing the first of two minicassettes Allan has talked full with briefs and letters. When that's done, I'll continue working my way through the dueling piles of books and papers he left on my desk. I'll reshelve the law books, organize and file client folders, sort through the mail.

The phone chirps. I pick it up on the first ring, simultaneously sliding open my desk drawer with my knee to retrieve a pen. "Good morning, law offices."

"Sue Bee!" It's my good friend Galen, the receptionist for the architectural firm next door. "Is it lunchtime yet?"

This is the third time she's called and asked the same question. It's eleven o'clock.

"I wish," I say, pulling the cap off my pen and coloring in the nicks on my desk. "I swear this place is a time warp. I mean, time slows to a third as soon as we get here."

"Don't I know it."

"Maybe we should call Stephen Hawking: 'Uh, Mr. Hawking? I think we've found a black hole.'"

Galen's phone is buzzing. "Hey, I gotta go. McCormick's for lunch?" she asks in a hurry.

"Mmm. Crab cakes." I drop the handset back into its cradle.

Galen and I work in Ghirardelli Square. One of San Francisco's biggest tourist attractions, the former chocolate factory is filled with tchotchke shops selling cable-car figurines and mass-produced watercolors of the Golden Gate Bridge. The offices where we work are on the second floor and look out onto Alcatraz. "From one prison to another," as Galen says.

I never planned to be twenty-eight and using my journalism degree to familiarize myself with the administrative aspects of law, but here I am reduced to fantasizing about being a writer. I haven't had much luck getting published, but then again, I haven't tried very hard. I spend most of my free time hanging out with Galen and other friends, listening to music and going to clubs.

It's January 1995, a couple of years before the dot-com

boom, and San Francisco is a magnet for underachieving college grads like me. All of my friends have their degrees but are also choosing the path of least resistance, working as waitresses, secretaries, and housepainters—dead-end jobs that pay maximum dough for minimum brain power. I'm heading into my fifth year as a secretary. Though the work is easy and pays okay, I don't want to make it my career. Miss Moneypenny? I don't think so.

I'd like to do something more with my life than answer phones on the first ring and type at the speed of human conversation. Since I can't get paid for writing, I figure I should volunteer for something else I'm interested in: radio. Despite growing up with parents who idolized John Denver and an older brother who wore out his eight-track playing Kansas, Boston, and Steely Dan, I love music, even if my white-bread suburban Chicago upbringing left me with some pretty severe musical gaps. I'm a huge fan of the Clash, but I'm only now hearing about the Minutemen. I've listened to James Brown for years but Parliament only recently. I'm learning these things thanks to college radio, which, for some reason, I'm only discovering now, years after graduating with my bachelor's.

KUSF is my favorite station. It's also the only college station I can get at my house because the signals for the others are too weak. Broadcast from the University of San Francisco, KUSF's DJs have turned me on to Royal Trux, Rapeman, and countless other bands I would never have known about had I been listening to any of the other stations in town, all of which are a disgrace not only to radio but to a city as progressive as San Francisco. On the right end of the dial, above 92 FM, are the commercial stations, where the playlists are about as long as my thumbnail and ads get as much airtime as the music. I'm particularly disgusted by Live 105. For a station that claims to be the place for "modern rock of the '90s," they sure play a lot of the Cure (which isn't modern), Dramarama (which isn't rock), and New Order (which isn't '90s). Then there's public radio, which I appreciate but don't want to listen to. The lefty politics and Celtic dance programs are like fingernails on a chalkboard for a motorcycle-riding blonde with a bunch of leather in her closet. I'm young. I want to rock. Is that really too much to ask from radio?

There has to be good music out there somewhere. There's got to be something more than Mariah Carey and the Spin Doctors. Bad radio seems like such an easy problem to fix. I don't understand why there's not a team of scientists working on the problem. Radio could be so much better. *I* could make radio better.

As much as I love college radio, I realize it could be improved, too. KUSF's music programming is great, but it stops at six every night and switches to a Chinese talk show. So, from six 'til midnight, there's a void. A void that needs to be filled. I've heard that FM licenses cost only a couple of thousand dollars, so I figure I should get one. I can't believe no one else has thought to do this. It seems so simple.

How I can buy a radio license when I can barely pay my Visa bill, I don't know. What makes me think I can program a station better than the supposed pros when I own only a dozen albums, I have no idea. But some combination of willpower, boredom, and disgust prompts me to meet with a lawyer and get the ball rolling.

Allan, in addition to representing real estate companies, takes clients through California Lawyers for the Arts—a referral service that hooks up poor countercultural types with altruistic attorneys. I call the group and request an attorney who specializes in radio. I'm paired with Peter Franck, a fiftysomething gray-hair with an impressive address in a downtown high-rise and a weakness for hippie causes.

I meet with him over an extended lunch break. After his receptionist leads me into his office, Peter shakes my hand and waves me toward a chair. Then he seats himself, pushing his glasses down his nose and peering at me over the rims.

"The form you filled out says you're interested in an FM license?" he asks.

I nod.

"What is it that you do for a living?

I tell him I'm a secretary.

He raises his eyebrows. "You must make good money."

I don't catch his joke and tell him my salary—$38,000.

"Are your parents wealthy? Do you come from money?"

I don't.

"What kind of assets do you have?"

I tell him about my motorcycle—a third-hand rice burner worth $500.

"I see," he says. He takes off his glasses, sets them on his desk, and gazes at me much like I imagine Christopher Columbus looked at the king of Spain before informing him the world was round.

You can't just buy a commercial FM license, he tells me. Sure, the application fee is only about $3,000, but licenses are renewed by the radio stations that already have them. That means you have to buy the radio station, and in San Francisco even a small station could cost as much as $20 million. The only way to get in on the cheap would be to apply for a frequency as soon as one opens up, but there's no possibility of that happening. Like all major U.S. cities, the FCC has already divvied up the FM dial into as many frequencies as possible, given the agency's current technological restrictions.

Suddenly I'm feeling like the dumb blonde I'm often mistaken for. I want to crawl out the window and onto the ledge, but instead I just sit there, crimson. He smiles warmly, sensing my embarrassment, and asks if I've heard of Stephen Dunifer. The name sounds vaguely familiar, but I ask him to fill me in.

Dunifer runs a pirate radio station called Free Radio Berkeley. He has been on the air since April 1993, even though the FCC busted him just a month after he started broadcasting by serving him with a Notice of Apparent Liability that levied a $20,000 fine against him. Instead of paying up or going off the air, Dunifer continued hiking into the Berkeley hills every Sunday night, carrying a transmitter, mixing board, microphone, CD player, tape deck, and battery in his backpack so he could set up shop in a tree and broadcast from there. During that time he was also working with a group of attorneys who helped him appeal the fine to FCC headquarters in D.C., raising the issue that its ban on low-power radio was unconstitutional. Dunifer's argument: The agency was violating his First Amendment right to free speech by denying him access to the airwaves. Instead of dismissing his case as frivolous, the judge ruled that his argument was serious and substantial and ordered the FCC to address the technological and constitutional issues raised in Dunifer's appeal. Dunifer knows the legal system well enough to understand that his case won't be decided for years and has taken advantage of the judicial lag time to jump-start an illegal-radio revolution. He designs his own transmitters and sells them to aspiring pirates around the country.

It turns out Peter Franck is a member of the National Lawyers Guild's Committee on Democratic Communications, the pro bono group of attorneys who are defending Dunifer from the evil clutches of the FCC and helping him radicalize the airwaves.

"I can't tell you to violate the law, but I do know Dunifer got on the air for very little money," Franck tells me.

I ask how much. It's about a thousand dollars.

Franck probably shouldn't be telling me this. Even if he's basing his case on the claim that it's the FCC, not Dunifer, that's acting illegally, as an attorney he should counsel me against engaging in an illegal activity that could land me behind bars for a year and lead to my wages being garnisheed through retirement. True, the FCC might not be regulating the airwaves in the public interest like it's supposed to, but Dunifer is still breaking the law because he doesn't have a license to operate. On some level that scares me, but fear is part of the draw. I'm attracted to pirate radio for the same reason I ride motorcycles: It terrifies me. In the half decade I've spent answering other people's telephones and watching the clock inch toward closing time, I've become bored.

I've always done what's come easily. In school, it was simple to pull A's and B's with little study, but when it came time to put my brain to use, I faltered. I'm a secretary, not a writer, because I type 160 words per minute—a freakish by-product of ten years of piano and violin lessons as a kid. Being a journalist would have meant too many sacrifices—moving to a small town, working for a tiny newspaper, making almost no money—and I didn't want to do that. It was too much work.

There's no obvious reason for my lack of motivation. I come from a professionally successful family. My dad's a nuclear physicist, my mom teaches English at my old high school, and my brother and sisters have each figured out what they want to do and are doing it. Then there's me, the administrative assistant.

My mother told me that as a child I never asked for what I wanted. If there were Oreos on top of the refrigerator or Barbies in the toy box, I'd just stare at them. For hours. My behavior was so unambitious, so different from my siblings, that she thought I was retarded.

It's taken almost five years in San Francisco for me to begin to figure out who I am and what I want to be. Call it delayed rebellion, but I like the idea of doing something illegal and getting away with it. I've been playing by the rules for too long, and it hasn't gotten me anywhere. Before I moved to San Francisco my definition of the good life involved a satisfying job, my own house, and a great boyfriend, but that dream is beginning to feel empty, probably because my job is dull, the two thousand dollars I've managed to save is far from buying a house, and don't even get me started on how some of the boyfriend stuff has worked out. The more I think about pirate radio, the more I like the idea of creating something unique and totally my own. I need to make a radical shift to get myself out of this rut I've been living in.

With the exception of what Franck just told me, all I know about pirate radio I learned from *Pump Up the Volume*, the 1990 movie starring Christian Slater as Happy Harry Hard On, the renegade teen DJ with a rocket in his pocket and an illegal transmitter in his basement who kept the kids entertained after school with an on-air mix of masturbatory improv and alternative rock while his unsuspecting parents puttered around upstairs. I loved the movie, but I never thought the ninety minutes I'd spent eating popcorn and watching Happy Harry try to outrun the FCC enforcers in his mom's Jeep had made any lasting impression. I never imagined Happy Harry would be me.

Yet here I am five years later with the phone number of a pirate radio guru in my backpack. When I call, I get an answering machine with a long-winded message telling me the FCC are motherfuckers and that corporations are fascist. "Micro power to the people!" the disembodied voice says, before instructing me to leave my name and number if I'm interested in seizing the airwaves and joining the revolution to free radio from the nefarious FCC.

I leave my number. Dunifer calls me back a few days later, at which time I rapid fire my most pressing questions: What does the FCC look like? Do they really drive white vans? What will happen to me if I get caught? Aren't you scared of going to jail? Where is your station? How much does a transmitter cost? Where can I get one? When can I get one? Can I come over now?

"What are you? A journalist?" he asks once I let him get a word in.

I get the feeling that reporters rank second on his shit list, just after the FCC, but I'm flattered. I've had only one byline.

"What do you want a station for?" he asks, a slight sneer in his voice.

Music, I say.

"Music?" he snorts. "There's enough of that out there already."

He's right. There is. What Dunifer doesn't seem to understand is that using micro radio to broadcast alternative music is just as necessary as it is for alternative politics, since most of what's aired is utterly awful, overplayed, or both. I suspect Dunifer's never heard the choreographed, cookie-cutter bop rock of Boyz II Men. He's unaware that a hit song like Nirvana's "Smells Like Teen Spirit" has been played almost ten thousand times in four years, and that's on a single radio station. Those are the problems I'd like to fix. There's a lot more music out there that deserves to be heard.

Still, Dunifer thinks my project is a waste of time and has no problem telling me so. When I reveal my plans to call the station KPBJ, he's even more incensed. Using four letters, the first of which is a *K*, buys into the FCC's system, he tells me. It

doesn't seem to register with him that by naming my station after a sandwich I'm also poking fun at the system.

He also takes issue with my use of "pirate" radio. "I don't
particularly like the term," he says. "To call yourself a pirate is to
admit to felonious activity and to buy into their language. What
we're doing is a constitutionally protected legal activity. It's just
that the FCC does not recognize the existence of the U.S. Con-
stitution."

I ask what language he'd prefer to use.

"Free radio, micro power broadcasting," he says. Grassroots
and community radio are also acceptable terms.

In his opinion, free radio is for serious political issues, which
to my mind amount to discussions on how to save the planet in
seven easy steps or feed the world through hydroponic garden-
ing. Though I classify myself as a liberal, politics is about the
last thing on my mind, mostly because I have absolutely no idea
what's going on in the world. I don't have a TV, don't listen to
NPR, and never read the newspaper. The news, or "blues" as a
friend calls it, is depressing, so I steer clear.

Squatting in the center of the FM dial without a license is
enough of a political statement for me, but in Dunifer's eyes my
plan to broadcast music is all about my ego—a "vanity station."
I'm tempted to tell him I'll call the station KSUE instead, just to
hear how quickly he'd hang up. I don't quite follow Dunifer's
logic or what makes his efforts any less vain. I mean, if this guy
is fighting for free speech, then shouldn't he be interested in
freedom of expression? I opt against a fight. It's hard to argue
with lefty liberals. Their minds are set like cement, and I don't
have a jackhammer. I cut to the chase: Can I buy a transmitter
from him or what?

"Yeah, sure," he says. "But you'll need to come over here. I
don't deliver. And you'll need to bring cash."

Like all rebels with a cause but no gainful employment, Dunifer is eager for money. I ask how much, and he tells me it will be $250 if I buy a kit, $350 if he puts it together for me.

That weekend I'm standing outside his live/work space with a stomach full of butterflies and a wallet stuffed with twenties fresh from the ATM. Dunifer's place is several blocks from the freeway on Allston Way, only one street off the main drag through Berkeley. I'm surprised such a covert operation is so centrally located. I question whether Dunifer gave me the correct address, especially after I take in the enormous mural of a radio dial on the front of the building. Then I realize the mural isn't his. It belongs to the legitimate radio-repair shop in the building.

I press the buzzer. Dunifer takes his time answering the door, but eventually he comes to get me. Except for his long gray hair, which is pulled into a low and loose ponytail, he reminds me of Pa Joad, wearing wire-rimmed glasses and baggy jeans with suspenders. I guess he's about forty, but he seems twice as old and moves half as fast as he should for his age. His physical frailty is a stark contrast to the bullheadedness of his politics. Later, I learn he suffers from arthritis, but at the time I don't know what to think.

After the cursory exchange of hellos, he leads me upstairs to show me the space where he builds the transmitters. It's kind of like Santa's workshop—if Santa's a skinny, humorless grump who lives in a dungeon. The wood paneling makes the space seem far smaller than it actually is, and there's no natural light—just a bunch of fluorescents hanging from the ceiling. About the only color in the room comes from the electronic parts and savaged computer carcasses whose wiry green guts spill onto the metal shelves where they are stored.

I'm only mildly interested in the tour. What I really want is to get my hands on a transmitter. I ask if he has one for me, but Dunifer says he mailed out his last kit that morning. I don't understand why he did that when he knew I was coming over to pick one up, and I don't have the guts to ask. I suspect the only reason he invited me over is to see what kind of person I am. At any rate, he tells me he'll have more parts in a few days, and we set up another time to get together. I keep my twenties in my wallet.

The following week I ride back to his house. When I ring the bell, no one comes to the door. I find a pay phone around the corner and call, but he doesn't pick up. I go back to his warehouse and buzz again. Nothing. Thinking he might be in the bathroom or talking on the phone, I wait around another fifteen minutes, periodically ringing the buzzer, but eventually I give up. Pissed, I ride full throttle back over the bridge to my house. He calls me later that night and apologizes, saying he forgot about our appointment. We set up a time to meet the following night.

Like a scene out of *Groundhog Day*, he never shows, only this time he doesn't return my calls. I don't know what to do, but being blown off only strengthens my resolve. If Dunifer wants to have a stubborn contest, he isn't going to win. I call Peter Franck to ask if there's anyone else I can get in touch with. As luck would have it, Franck knows about another local pirate who was raided and is thumbing his nose at the FCC by staying on the air. His name is Richard Edmondson.

Richard went on the air just a couple of weeks after Dunifer, broadcasting from a mobile home on top of some of the city's highest hills—including Portrero Hill, where the FCC eventually tracked him down. For the past year he's been running San Francisco Liberation Radio from his one-bedroom apartment in the city's Richmond district, where there's an abundance of fog and Chinese restaurants but limited listenership for the politics he's broadcasting. Richard's station, like most micro radio broadcasts, is too low power to reach outside his neighborhood. San Francisco's multitudinous hills block the transmission. I've tried, but I haven't been able to tune in to SFLR, and I'm only two miles away. Richard is extremely friendly when I call. Without hesitation, he invites me over to check out his operation.

Even if I hadn't had the address, I would have been able to figure out which apartment was his from the towering, twenty-five-foot antenna on the roof. Richard's station is set up in his living room, the housing equivalent of an overstuffed suitcase. The space is jammed with tape decks, mounds of cassettes, CDs, CD players, records, record players, microphones, microphone stands, audio cables, plastic lawn furniture, and broadcast gear. Richard's girlfriend Jo, a preschool teacher, is on the air when I get there, reading letters supporting convicted cop killer Mumia Abu-Jamal. They ask if I'd like to help read them over the air, and I say yes even though I have no idea who Mumia Abu-Jamal even is. He hands me a stack of letters, sits me in a chair, and passes me the microphone.

As I read, I wonder who's listening. I imagine it's disgruntled Unabomber types who have their Walkmans tuned to SFLR as they wire together mail bombs or plot to kidnap the mayor. It's more likely that the three of us are the only audience for these recitations, but that doesn't matter. I'm on the air, getting my first rush from an illegal broadcast.

To give us a chance to talk, Richard fishes around in an old cardboard box filled with cassettes of prerecorded political talk programs from fellow micro radio operators around the country. There are cassettes for a media watchdog show called *Counter Spin*, the punk rock program *Maximumrocknroll*, and *Copwatch*, which catalogs police brutality in Berkeley. Without even looking to see what it's about, he loads one into the tape deck and presses Play.

Living in San Francisco, I'm surrounded by political activism of every kind, but it isn't until I visit Dunifer's workshop and Richard's apartment that I get a firsthand, up-close look at this sort of ultrapolitical lifestyle. It's so far outside my reality. I'm just a twentysomething slacker living under the shadow of parental expectation. They're radical activists battling for what they believe in. They have the courage to stand up and fight, and they're willing to suffer the consequences—a diminished standard of living, police raids, jail time. I'm impressed with their dedication and spirit, but that's them. I'm not ready to sacrifice my creature comforts for a cause just yet.

I guess that Richard is about thirty-five. Like Dunifer, he doesn't have a job. He gets by on disability payments. Government assistance, it seems, is inadvertently helping to fund micro radio. Richard keeps SFLR on the air for twelve hours every day. Starting at four o' clock in the afternoon, Richard or someone else from the community deejays live—talking politics, playing whatever music was mailed to the station for free, and interviewing activists involved with any and every cause, from feeding the poor as part of the Food Not Bombs campaign to establishing rights for the homeless. Around 10 P.M., he retires the mike, but SFLR stays on the air replaying tapes until 4 A.M.

Richard's programming is vastly different from what I imagine doing with my station. I want to take the best music from college radio—stuff like PJ Harvey or Portishead—and form it into a loose playlist that lets the DJs have control over what they play but also ensures that a certain number of new songs air every hour. I presume that, like me, a lot of listeners want to hear what's just come out. My friends will be the DJs, and they will broadcast only a couple of hours each night, but all the programs will be live. We'll have some fun, and listeners will get to hear some great music. Everyone wins.

Because Richard is so much more political than me, I'm not sure how he'll react when I tell him my plans, but he's extremely supportive. When I tell him the trouble I'm having with Dunifer not returning my calls, he encourages me to call Chris A.—an unpaid but politically aligned protégé of Dunifer's who has access to the micro radio guru's secret stash of transmission gear. Chris helped build Richard's transmitter and everything else he needed to get SFLR off the ground.

I call Chris as soon as I get home.

He answers on the first ring. "Hello?" a voice says suspiciously.

"Hi. Is Chris around?"

"Yes."

"Can I talk to him?"

"Yes."

"Are *you* Chris?"

"Yes."

Social skills are frequently inverse to technical knowledge, I remind myself. I learned that lesson long ago, having grown up with a scientist for a father. I introduce myself and drop Richard's name as if it were the password to a speakeasy. It seems to do the trick. When I ask Chris if he can help me get a transmitter, he gets a kit for me the following week. He even invites me over to help build it.

I go to Chris's house one night after work. He lives in the Tenderloin, a part of San Francisco that has almost as many people living on the street as indoors. In my quest for a transmitter, I'm getting used to showing up on strangers' doorsteps in questionable parts of town as if I were a crack addict looking for my fix.

It's not like my friends' apartments are a whole lot better, but at least I know what I'm in for when I step inside them.

Chris lives in a rundown Victorian row house so severely damaged in the 1989 earthquake that it's angled and cockeyed. I ring the bell, he buzzes me in, and I climb up to his third-floor flat on stairs that are like a fun house minus the mirrors. As I approach the landing, I see Chris suspending himself in midair, holding on to the handrails.

"Hi," he says, leaping down onto the step in front of me and extending a hand. "I'm Chris."

From his staccato phone manner, I'd imagined Chris might wear a pocket protector, but I was way off. He's quite a bit younger than me, twenty maybe—a stickpin of a guy with ruddy pink skin and curly red hair that seldom sees a brush.

He leads me past a trio of mountain bikes in the hallway and into his living room, which is furnished with nothing but a couch, a lamp, and an old cardboard box he uses as an end table. Like Stephen Dunifer and Richard Edmondson, Chris doesn't have a job. He's one of those idealistic types who get by on nothing while devoting all his time to the causes he crusades for—feeding the poor, disarming the government, freeing the airwaves, and adding bicycle lanes to city streets. Chris tells me his causes have led to his being arrested several times, at least once for attempting to ride his bike over the Bay Bridge during rush hour.

Fishing around in a black Hefty bag parked against the living room wall, he extracts a blueberry bagel for himself and offers me one. I decline, even after he holds the bag open to show me it isn't filled with trash but breads from a local bakery. Tomorrow, he'll wheel around the city on his bike delivering them to the homeless, he says.

Chris invites me to have a seat on the couch, after giving it a few ineffective sweeps with his hand to brush off the crumbs. I trade him my money for the transmitter kit. It's in an unmarked, nine-by-twelve manila envelope. Opening it, I find a circuit board about the size of a CD jewel case and a sandwich Baggie filled with minuscule multicolored pieces that look like beads with prongs. The instruction sheet appears to be written in English but reads like Vietnamese. I know a tiny bit about mechanics from working on my bike, but this is far more complicated than any motorcycle service manual. I'm tempted to quit before I even begin.

When Chris invited me over, I'd secretly hoped he would put the transmitter together for me—that I could just hover over his shoulder, contributing nothing more than an occasional nod or smile of encouragement. No such luck. After plugging in what looks like an electric screwdriver—it's a soldering iron—he picks up a book and hits the couch.

Chris doesn't have a table, so I sit on the floor. I dump what's inside the envelope onto the carpet. What a mistake. Every time I reach for one of the pronged thingamajigs, my hand comes into contact with shriveled black beans, clumps of mud, bits of wire, grass, or hair.

I don't want to touch the floor, and I don't really know what I'm doing, so I'm not making a lot of progress. After a few minutes, I catch Chris peeking over the edge of his book. I suspect he hasn't even been reading—that he knew it was just a matter of time before I'd play the damsel in distress. He climbs off the couch and helps me sort out where everything goes, then watches me solder a few pieces in place to make sure I'm doing it right. I am, sort of.

Chris runs a finger over the circuit board's underbelly, where I've left mountains of lead while installing the electronic chips. When soldering, I've been applying the same philosophy

I have for conditioning my hair: More is better. He suggests less is best and leaves me alone. It takes me four more hours to finish the job.

Gloopy is the word Chris uses to describe my finished product. He doubts that the transmitter will work but says he'll take it to Dunifer's workshop to test it. Exhausted from the technical effort and delirious from the lead fumes, I go home. It's close to midnight, and I've got to work tomorrow.

As Chris suspected, my transmitter doesn't function properly, so he swaps it for one Dunifer had assembled himself. I think the one I made is shipped off to another aspiring pirate. My transmitter is just the first step to getting on the air, though. I need an antenna and audio cable and a never-ending list of nickel-and-dime stuff: a wire cutter to strip the ends of the cable, BNC adapters to fit on their ends, a tuning wand to tune the transmitter, a power meter to measure how much juice we're getting out of the system, a fan to keep everything cool, a power amplifier, more solder. Happy Harry had just flipped a switch, thrown on a Pixies record, and called it a done deal. He didn't solder or make daily field trips to Radio Shack. *Pump Up the Volume* didn't show all the shit work that's involved in building a radio station by hand. If it had, it might have been unwatchable, and I might not have continued with my plans.

My transmitter is 20 watts and uses as much energy as a lightbulb with the same rating. At such low power, I expect the signal to reach about a mile, but how far it travels depends on the design and placement of the antenna. I could buy an antenna at a ham radio outlet, but it would set me back a couple of hundred bucks. Chris offers to build me one for half as much, using copper pipes he can buy from the plumbing department of any hardware store.

I also buy a microphone and a mixing board from Radio

Shack—my new favorite store. They cost about a hundred dollars each. Like Richard Edmondson, I plan to reconfigure the components of my home stereo system—the CD player, turntable, and tape deck—and run them through the mixer for my jury-rigged version of how the real radio stations do it. The only thing I'm missing is the table so I can lay it all out. To save money, I build one myself. I could buy one, but it needs to fit in a very specific and limited space—the narrow pathway between my bathroom and bed.

My table has eleven legs and sways when I lean on it. This isn't such a good sign, considering the thousand dollars' worth of stereo equipment I intend to pile on top of it. Imagining the table crumbling under the weight of my record player, mixing board, tape deck, and CD player, I push the wobbly monster into the corner and nail it to the wall. Problem solved.

The apartment that will soon double as KPBJ is New York–sized—a tiny studio tucked at the back of a three-bedroom house that is shared by three people: a waiter for the Ritz-Carlton, a handbag designer for Esprit, and a skydiver. Technically, they aren't my roommates. Our living spaces are separate. My apartment has its own kitchen and bathroom and a separate door that is accessed through the backyard. Now that I've started building the station, I'm not sure I should tell them. I've invested too much money, time, and energy in the project to give it up if they object, but my conscience gets the best of me. I'm about to fasten a rather conspicuous twenty-foot pole to the roof. If the landlord stops by and makes a stink about it, we'd better have our stories straight or all of us could be evicted. I decide to spill the beans.

I don't know why I'm so worried. Running an illegal radio station is about as risky as boiling water to someone like Bill, who commutes two hours north of the city to stuff parachutes and jump out of planes. During his free time, he bungee jumps from bridges and water towers and parachutes from the forty-second floor of the San Francisco Hilton.

"Cool" is all he has to say when I reveal my big plans, and our two housemates feel the same.

My apartment is near the top of one of San Francisco's finest hills—the monumentally steep Sixteenth Street, which overlooks the Castro, the Mission, downtown, and the Bay. Despite the height and spectacular view, my location isn't ideal for radio broadcasting. My house is slightly below the crest of the hill, and FM radio is line of sight. The signal reaches only as high as the antenna, and my antenna isn't tall enough to transmit over the hill and into the Haight-Ashbury district where my target audience lives. My signal will reach only those people who populate the neighborhoods I can see from my roof.

I could have built some impossibly complicated and dangerously tall antenna tower in my backyard, but, given the transmitter and table I've made, such an antenna would be just as likely to fall into a power line and electrocute my entire block as it would be to radiate a signal over the top of my street and toward the rockers, ravers, hipsters, and hippies I hope will listen to my station.

Who will be listening is the least of my concerns, though. First, I need to get on the air.

CHAPTER

It is late June, almost six months after I first contacted Dunifer, when Chris and I make a date to install the antenna and begin broadcasting. As a thank you, I make him dinner. I figure the kid could use a decent meal since bagels appear to be his only food source. It's a Saturday, late afternoon, when he arrives, slick with sweat from an unsuccessful attempt at pedaling up the near-vertical hill to my house. I extend my usual greeting—a glass of water—and continue rolling meatballs—meatballs he can't eat because he's vegan. Some thank you, trying to feed cow to a vegetarian, but Chris just smiles. He picks up the un-broken round of sourdough on my counter and chews through the entire loaf in ten minutes, talking logistics between bites and sips of water. We walk outside.

The antenna needs to make it onto my roof, but there's no easy way to get up there. I don't have a ladder, and neither of us is a Chinese acrobat. We scout the perimeter of the house, look-ing for something to climb. There's a drainage pipe attached to the garage, but the garage is not attached to my house. There's a six-foot gap between the buildings. Chris scales the pipe anyway. He jogs across the shingles of the sloped roof, gathering as much speed as possible. He leaps. He flails. He lands in a clump on my roof. I applaud. He brushes the fiberglass from his palms, then he bows.

I've got the easy job—handing him the antenna and electrical cable. Chris drops the pole into a vent hole. I have no idea what the vent is for, but its width is a near-perfect match for the pole, so I don't care. The antenna will be stable. That's what matters. In a spine-expanding stretch, I reach the cable up to Chris so he can connect his end to the antenna. I unspool the rest, snaking it through a hole I rip in my window screen and under my futon to the little silver box that houses the transmitter. It's cold at night and I don't have heat, so I'm not so happy that the window above my bed will be cracked at all times, but it's the shortest route from the transmitter to the antenna. And the shorter the cable, the less likely it is we'll lose precious wattage in the transmission. A radio station or my health? I'll take the radio station.

Chris jumps from the roof and comes inside. I've already set up the "studio" on the multilimbed table I've built and nailed to the wall outside the bathroom. All that's left to do is tune the transmitter to 96.9. In the Bay Area there are a handful of FM frequencies that can be used for low-power radio because they are spaced far enough away from other stations on the dial that they won't cause interference. The other frequencies are already being used—by Dunifer, Edmondson, a militant Latino political station in the Mission called Radio Libre, even Chris A., who runs a little station from his cockeyed apartment in the Tenderloin. With the exception of Chris, who rarely goes on air, all of them have been caught and are waging war with the FCC.

I can't help but wonder how long it will be before the men in blue windbreakers stop by for an impromptu interrogation, but it's too late to turn back now. KPBJ will be broadcasting within an hour.

Sitting on the edge of my futon, Chris unscrews and removes the lid of the transmitter. The station is transmitting. The Jon Spencer Blues Explosion is playing on CD, but I can't hear it. The studio receiver is tuned to 96.9 and the volume is up, but instead of hearing Spencer's campy yowl, all that's coming from the speakers is a weird whistling noise. Its shrillness depends on what components Chris is messing with inside the transmitter. Using a tool that looks like a pen crossbred with a screwdriver, he spins some tiny components inside the little silver box. The whistling quiets down only when he locks on to the frequency.

That's when I hear it: "Bellbottoms!" Jon Spencer is coming through loud and clear on 96.9. "Bellbottoms," Spencer spits into the mike. "They're gonna make you wanna dance!" I scream, laugh, and nearly trip over the transmitter as I scramble over to Chris and give him a hug. I am out-of-my-mind ecstatic. But I'm also terrified and worried. I am now an outlaw. Oh my god. I could go to jail.

The criminal penalties for operating an illegal radio station are up to a hundred thousand dollars in fines and one year in jail. I don't have that kind of cash, and I'm not at all interested in a wardrobe that consists of one outfit: an orange jumpsuit. No. If I get busted, I want to protect Sue Carpenter, and pretending to be Paige Jarrett seems the best way to do it.

Paige Jarrett is a girl I knew in high school, a girl I wasn't even friends with and knew almost nothing about except that she was on flag corps and drove a vintage, cherry-red Mustang that looked as if she Turtle Waxed it before bed every night. We didn't know each other well, though we might have made eye contact in trigonometry class or essay writing. It's possible we even said hi a few times as we passed herdlike through the echoey hallways of Naperville Central, in a Chicago suburb, but we were not friends. Still, we were a lot alike—shy and kind of nerdy. She just had a better name, so I borrowed it.

Paige and I have enough in common. She's twenty-eight years old, five-foot-eight, 135 pounds, blond, and green eyed, at least according to the stats I copy off my driver's license and dump into the fake ID card I dummy together at my office. For the picture, I use an old Polaroid of myself. For the address, I list my work. Then I take the whole mess to a copy shop and have it laminated. My new ID card looks nothing like any ID I've ever seen before. I just hope it works.

I figure I'll need it to set up voice mail and a mailbox in Paige Jarrett's name. I don't want to use a live phone line for the station because it would be too easy to trace the number. I want the mailbox service so I can get free music from record labels, but I want it delivered someplace other than my house. The ID, it turns out, is unnecessary for either. Still, I hang on to Paige's name so she can contact record companies and ask for music.

After KPBJ has an address and a phone number, I design the letterhead—the logo is a photograph of a peanut butter and jelly sandwich—and have it professionally printed. Then I use the stationery to write letters to about fifty independent record labels, the names and addresses of which I find in the back of the music 'zine *Fizz*. The letters introduce KPBJ as "a tasty alternative to college radio" and ask for their support in the form of free CDs and records.

I don't expect many labels to respond, so it's several weeks

before I check the mailbox. I go there after work one evening
and unlock the box. In it is a slip asking me to go to the pickup
counter. I step up to the desk and hand the square of paper to
the man behind the counter. He looks at my mailbox number
and then at my helmet.

"You're on a bike?" he asks, before disappearing into a storage area.

I think he's just making conversation, but the real reason for
the question is soon apparent. He comes back to the counter
with an armload of manila padded mailers—more than I can
fit in my backpack. I can't believe so many of the labels sent me
music when they don't even know me or the station. They don't
even know if it really exists. Small record labels are, obviously,
starved for airplay.

I put as many of the packages in my backpack as will fit, stuff
the rest inside my jacket, and ride home feeling like the Michelin Man—a Michelin Man who's just won the lottery.

At home I rip open the packages from Tim/Kerr, Epitaph,
Kill Rock Stars, Dischord, Touch & Go, Drag City—record la-
bels I'm only marginally familiar with. It's time to buy a new
CD rack.

I barely know anything about these labels, let alone the
bands—artists like Elliott Smith, the Red Aunts, the Dandy
Warhols—but I'm still ecstatic. I spend the rest of the night listening to them, making notes that describe their sound and
picking out my favorite tracks. Then I write the information on
stickers and stick them to the jewel cases.

I make a list of all the CDs I received and send thank-you
letters and the KPBJ T-shirts I've had made to all the labels that
sent me music. I figure, if they're giving me something, I should
give them something back. Most of the people who get the
shirts seem pretty surprised. Either that or they're horrified.

The design for the T-shirt is a sandwich, with KPBJ written in peanut butter and jelly between the two slices of bread. Abby, who's one of the DJs, designed the logo, and I think it's cute, but other people think it looks like a dirty menstrual pad, so who knows.

I have some, but not much, experience as a DJ. I know enough to preview a record without playing it on top of what's already going out over the air, but that's pretty much it. I have no on-air personality to speak of, and even less knowledge about what I'm playing. In a nutshell, I suck, but what little I do know I learned as a volunteer for the area's two college radio stations— KUSF and KALX, at UC Berkeley.

A couple years before pirate radio had even registered on my radar, I'd joined KUSF as a volunteer. From the get go, I was doomed. First, there weren't that many DJ slots available. Second, the people who had them had been on the air for years and weren't about to give them up. And third, wannabe DJs outnumbered them twenty to one. With such dismal odds, I should've split after the first meeting, but I didn't. I signed up with the publicity department, thinking I could win them over through hard work. No such luck. In six months, I saw the inside of the studio only once, so I quit. I had no idea "community" radio would be so uncommunitylike.

Galen, my best friend and fellow secretary, suggested we go across the Bay to KALX, at UC Berkeley. A fellow music fan, she was also interested in deejaying. Galen had been to the station once, for an interview about a self-defense group she was involved with. She called the woman who had interviewed her and arranged for us to attend a women's department meeting.

I didn't know anything about KALX—just that I couldn't

tune it in at my house and that getting there involved a cold, half-hour ride across the Bay Bridge. I was also nervous about joining the women's department. I wanted to spin music, not talk about gender politics. I considered myself an individualist, not a feminist, and certainly not a Berkeley feminist, which, I presumed, meant hairy, lesbian man-haters who did nothing but read Gertrude Stein and march in pro-choice rallies. No, I wasn't one of "those." Those people scared me. Despite my reservations, it was our only prospect to get on the air, so a few weeks later, Galen was on the back of my bike and we were riding over the water to the well-worn studios of KALX.

I'd never spent much time in Berkeley, but KALX lived up to its hippie stereotype. Inside the station's waiting area were two couches that appeared to have been retrieved during a Dumpster dive. They were so stained that their only discernible color was dinge. The lighting was dim. Aging rock posters papered the walls.

There were only six women at the meeting, and that included Galen, myself, and Courtney Heller, the smoky-voiced brunette who was leading us. Clearly, I wasn't the only person with an aversion to feminism. Thirty thousand students and less than a dozen are interested in women's programming? At Berkeley? I'm stunned, but I see an opportunity.

The meeting was in a run-down production studio that was home to a monstrous mixing board, a couple of reel-to-reel tape machines, and a bunch of mismatched and creaky office chairs, where we all sat in a circle. Within the hour, Galen and I had learned how to operate both the mixing board and the tape machine. We signed on as volunteers that night, and, one week later, I was doing my first on-air interview. Six months later, I was running the radio station's women's department. Through volunteering at the station, Galen and I were eventually trained as DJs.

Galen is the only person at KALX who knows about KPBJ. If I told anyone else, I'm afraid they'd yank the student FCC license I'd gotten when I first signed on. But that doesn't stop me from using the KALX library to beef up the music selection at KPBJ. A lot of labels are sending me music, but it still doesn't seem like enough.

At KALX, we aren't allowed to sign out records from the library, but there are no rules to stop us from recording them. Once a week I sign up for the production studio in the middle of the night, when only the DJ and some other lost soul without a social life are around. From midnight to 6 A.M. I preview new CDs and records that KALX has gotten and KPBJ hasn't. If there are songs I like, I record them onto five-minute cassettes. My goal is to record thirty cassettes a night—about as many as I can fit in my backpack—then leave by 6:30, so I can beat rush hour traffic. I usually get home around 7 A.M. I set my alarm for one hour later and am answering phones at the law office by 9.

The second time I turn on the transmitter, I'm just as anxious, giddy, and terrified as I was the day before, when Chris and I first got the station on the air. I still can't believe I'm doing this. Setting up the station was hard, but running it seems too easy. You just flip a switch, and suddenly the airwaves are open to whatever you want to play. Unbelievable.

It's a Sunday night, and I've invited my ex-boyfriend Will to deejay with me. Actually, my ex-ex-boyfriend. We just got back together after a ten-month breakup—a breakup that was prompted by his budding love affair with heroin and flirtations with other girls. While we were apart, I cried more than I'd cried in all my twenty-seven years combined. I was so depressed that I threw myself headfirst into everything as a distraction from the pain. Those ten months were the most agonizing but also the most productive of my life.

I not only got the radio station together but upped the ante on my radio work at KALX, stepping up the pace of my interviews. Every week I read at least one, if not two, books, prepping for interviews with avant-garde novelist Kathy Acker, intellectual art critic Camille Paglia, pro-choice activist Jane Roe, and anyone else with a feminist bent. I also started writing. After chasing down a couple of meter maids on the street, I wrote a story about them for the *SF Weekly*. It was my first byline. And it was followed by other stories for the paper—about crank calls to crisis lines and the students at San Francisco's College of Mortuary Science.

So much progress, then Will steps back into my life. And now we're spinning records together. He adopts the on-air persona "Q"—as in James Bond's gadget guy. I am, wink-wink, Paige. Will's never been a DJ before, but he's in a band and knows a lot about music, mainly classical and rock. I show him how to use the equipment and we tag team at the controls. While the music is playing, we're making out like a pair of reunited love birds.

Kicking off KPBJ's second official night on air, I spin a set of PJ Harvey, the Red Aunts, Slant 6, Six Finger Satellite, and the song I once considered my personal anthem: Superchunk's "Slack Motherfucker." Will plays Black Flag, Sonic Youth, and the Soft Boys. Back to me. I play Helmet, Jesus Lizard, and Doo Rag. Back to Will: Joy Division, Tom Waits, and Boss Hog. We deejay for about three hours.

Unlike Will, my gift for gab runs in reverse, so my mike breaks are short and sweet. I just list the songs I played, then say the name of the station and the phone number. Since we've only just begun broadcasting, I'm fairly sure no one is listening, but I check the voice mail anyway. I punch in the code. There's one message. A guy named Troy called to say he's visiting from Orange County and just happened to find the station.

I save his voice mail. When I wake up tomorrow, I want to check it again and make sure this is real.

Will and I tag team at the controls again the following night. I'm playing Cibo Matto's "Birthday Cake" when a low-level hum begins to overrun the music. It gets even louder when I segue into Beat Happening's "Bewitched."

"Sweetie?" Will calls from the kitchen. "What just happened to the station? All of the sudden it sounds like shit."

"I'm trying to figure that out," I say, checking the levels on the mixing board to make sure I'm not overmodulating. Everything appears to be okay, but the hum is getting louder.

"What *is* that?" Will says, stepping up beside me and taking a long drag off his cigarette. "It sounds like a pack of killer bees."

"Or a buzz saw." I turn down the volume on the overall output, thinking that might be the problem. It doesn't help. "Shit. I have no idea what's going on here."

When Calvin Johnson stops singing, I get on the mike. "And that was Beat Happening, with some other annoying noise we can't figure out. Thanks for listening to KPBJ. We know it sounds like total crap right now, but bear with us."

Will leans into the mike. "And if you have any idea how to fix a pirate radio station—we sure as hell don't—give us a call. Here's some Nick Cave."

I turn off the mike, and Will pushes up the fader on "Do You Love Me."

I try calling Chris A. He isn't home. I rejoin Will and continue with the music, hoping the problem will clear up on its own.

While one of the songs is playing and I'm cuing up a tape, Will checks the voice mail.

"Sue," he says, hanging up the phone. "Some guy just called to say he knows how to fix that weird noise."

"Did he sound legit?" We've only been on the air a few days, and I don't want to fall into any traps.

"I don't think it's the FCC, if that's what you're thinking."

"Of course that's what I'm thinking, you freak!"

"Sue. Do you really think the FCC would call you first? I think they'd just crash through the wall." He cocks an ear toward the door, brown eyes sparkling with mischief. "Wait one second. I think I hear something."

"Are you making fun of me?"

"Never," he says, giving me a kiss on the cheek. "I saved the message, if you want to listen to it yourself."

"Did he leave his number?"

Will is one step ahead of me. He hands me an old envelope with his blue-ink chicken scratch. "He said his name is Larry or something."

"I don't know. Do you think I should call him?"

"What else are you gonna do?" Will takes the headphones out of my hands and points me toward the phone.

I'm nervous. I have no idea who this person is, but I dial the number anyway.

"Hello?" The voice on the other end of the line is a soft and husky whisper.

"Uh, hi, Larry? This is, um, Paige from KPBJ."

"Yeah, Chris told me he'd helped get a new station on 96.9."

"Chris? You know him?"

"Little guy? Red hair? Oh, yeah. He helped me out with an antenna for the little micro watt station I'm running, but we can talk about that later. You're running Dunifer gear over there, aren't you?"

"Yeah, how can you tell?"

"That hum is a Dunifer hum."

I laugh.

"Where are your power cords?" he asks.

"What do you mean?"

"The cords for your CD player and mixing board and the like? Where are they?" Larry asks, speaking slowly, methodically.

"Plugged into a power strip under the DJ table."

"Next to the transmitter?"

"Uh-huh."

"That's your problem. You've got to separate the power cords and get them away from the little hot box. They cause interference with the system. Go ahead and move the power cords."

"Okay, but I've got to put the phone down. It won't reach."

I move the clump of black cords away from the transmitter. The buzz completely goes away. I run back to the phone.

"Wow! It's that easy?"

"The cords induce an AC hum into the transmitter, which induces RF back into the audio."

"I guess so, even though I have no idea what that means."

Larry chuckles. "You do have a filter on your system, don't you?"

"A filter?"

"A little silver gizmo. It should be attached to the cable that runs out of the transmitter."

"I don't think so. Why?"

"You need to get one because your station might be broadcasting out of band into frequencies used by the police and air traffic control. You haven't had any trouble with the boys in blue, have you?"

"The FCC? No."

"You need to get a filter on your system ASAP."

He offers to pick one up for me over the weekend and hook it up. We arrange a time. I give him my address, thank him again and again, and hang up.

Will is deejaying. I join him. The night is young. There are records to spin.

With the exception of various friends who've stopped by to hang out, Will and I have been the only DJs all week, going on the air at 8 at night and playing music until at least 10. Ideally, I'd like the station to be on the air four hours every night, from 8 until midnight, seven days a week. Going on the air at 8 will give me enough time to have dinner and unwind after work. I'm hoping the regular hours will also help to build an audience.

Each shift is two hours long—just like at KALX. That means there will be fourteen shifts per week. There should be a different DJ for each shift, but I don't know that many people. Sure, I *know* a lot of people, but I'm friends with just a handful. And I only want friends to deejay. They're the only people I can trust.

I recruit Galen, her husband, Danny (a hockey-playing Canadian punk), her roommate, Abby (an Internet designer with more than a passing interest in aliens), Fire (Abby's boyfriend), Brad (a guy I had dated briefly who was also lead singer to a Ted Nugent and Led Zeppelin cover band called Ted Zeppelin), and Will, who just moved back in.

All of them are at my house for a meeting a week after we've been on air. I show them how to turn the transmitter on and off, and teach everyone who hasn't deejayed before (which is pretty much everyone) how to use the mixing board. We decide who will DJ when and discuss the station's format. I tell them they can play whatever music they want, but we should decide how many new songs should be played every hour. Once again, I'm taking my cues from KUSF and KALX, both of which require that DJs play a set number of new songs per hour.

Fire pipes up: "If this is a pirate radio station, why are we following college radio rules?"

He's got a point. Why are we? Because it's what I know. As I navigate the new and foreign terrain that is micro radio, I'm looking for guideposts wherever I can find them, and those posts are mostly at KALX. But Fire is right. This is illegal radio. There shouldn't be any rules, at least about what DJs can and cannot play. KPBJ isn't college radio. I change my mind: KBPJ will be an entirely free-format station.

The meeting is winding down when Larry, the guy who cured KPBJ's hum problem, comes over to add the filter to the transmission system. Why does no one ever look like his voice? I'd expected Larry to be gray-haired and somewhat professorial, but he isn't. He is overweight and older than me—by about a dozen years—but there's no gray hair, and he isn't wearing a tweed suit with suede elbow patches. He's dressed in a thin, button-down beige shirt, corduroys, and a backpack.

Like everyone else who accepts the physical challenge of visiting me, Larry is heaving and soaked with sweat when he arrives. I'm tempted to give him a towel. Sweat is beading on his blondish and balding head faster than he can wipe it away with his sleeve.

Drying a hand on his pants leg, he reaches out for a handshake but mumbles and looks away when he introduces himself. He attempts a smile, but his mouth doesn't fully cooperate. My heart immediately goes out to him. I've always had a soft spot for the socially awkward. I see more than a little of myself in such people, having been the requisite class nerd in junior high and high school—the skinny girl who walked the hallways hugging her foot-tall stack of books to her chest and avoiding eye contact. The girl who looked embarrassed to be alive: That was me.

I invite Larry inside and introduce him to Fire—whose leather-cuffed wrist is resting on the shoulder of his girlfriend, Abby—and Brad, whose facial hair is shaved into some unusual configuration involving a soul patch and sideburns. He meets Galen, whose long auburn hair is highlighted with skunk stripes, and Danny, dressed head to toe in black. Then there's Will, in his thrift-store polyester, and me, looking like the love child of Gene Simmons and Ani DiFranco, with my blond braids, jodhpurs, and motorcycle boots.

Larry is clearly a square peg in my hipster circle. From the way he is shifting his weight from foot to foot, I can tell he is extremely uncomfortable. In an attempt to make him feel at home, I offer him one of the peanut butter and jelly sandwiches I made for the meeting. He takes one from the plate. It is gobbled and gone in a mere three bites. I'm not sure if it's the observance of this spectacle that prompts it, but within a few minutes all of the DJs are getting up to leave.

Things are only slightly less awkward when Larry and I are one-on-one. To ease the discomfort, I slip into reporter mode, pummeling the poor guy with questions. I soon learn there's quite a bit of common ground between us. We both grew up in Illinois. Both of us ride motorcycles. And, obviously, we're both illegal-radio operators.

Larry tells me he's been a radio hobbyist since he was sixteen, when he built his own transmitter and began broadcasting from his bedroom. Using his sister's and mother's records, he played DJ, spinning Joe Cocker, Arlo Guthrie, the Temptations. Sometimes he recorded another radio station on his reel-to-reel tape machine, splicing out the announcer's voice and replacing it with his own. These days he works at a computer repair shop south of Market, just a few blocks away from Chris A. That's where he built the 3-watt transmitter he uses to broadcast from his apartment in the lower Haight.

Like me, Larry's apartment building is on a downward slope. Because of the geography and the low power he's broadcasting at, his signal doesn't carry very far or very well. He's less than a mile away, but I can't tune him in because his signal is weak and his antenna is lower than my house. My transmitter broadcasts with more power, and my antenna is higher than his apartment, so he has no problem hearing me.

Larry's been on the air for eight months. Like mine, his station is all music, not politics. Unlike everyone else I've met in the scene, he has not been caught by the FCC, and that's because he's extremely cautious. He operates at exceedingly low power and does not advertise. He doesn't even use his own voice when he's deejaying, fearing the FCC could trace the station to his house, figure out his phone number, match the voice on his answering machine to the voice on air, and bust him. To make announcements, he uses a computer voice program.

When Larry adds the filter to KPBJ's system, it is quick. The whole job takes five minutes. At least it would've taken five minutes if Larry hadn't felt compelled to open up the transmitter box and poke around inside. Techies. They simply can't control themselves around electronics.

Larry asks if I'd like to upgrade my signal from mono to stereo. Not only will KPBJ sound better, he says, it might help the station attract more listeners because some stereo receivers stop only on stereo signals. It might also protect me from the FCC people, who, he says, regularly cruise the FM dial listening for mono signals—a telltale sign the signal is pirate.

I ask how soon he can hook it up. He offers to come back in a couple of days.

* * *

The DJs begin their shifts the following week. Brad comes in toting bags of secondhand vinyl and CDs by Buck Owens, Mel Torme, Kay Starr, Tom Jones, the Ohio Players, and Captain Beefheart. Fire and Abby's taste tends toward college rock—Bikini Kill, the Pixies, Steel Pole Bathtub, Jesus Lizard, and Orange 9mm—whereas Danny and Galen stick to vintage punk like the Bags, the Fall, X, the Cramps.

We have some technical difficulties for the first couple of weeks, while the new DJs figure things out. Some of them forget to turn the microphone on when they talk, or they forget to turn it off so their conversations are broadcast along with the music. Others inadvertently play CDs at the same time they're using the turntable, or play everything so loud that the signal distorts and hisses.

We're beginning to get some calls—about one per show, though there are some nights we get zero. Most of them are requests, for the Beastie Boys, Al Green, Jerry Garcia, John Coltrane, the Partridge Family, the Cows. There are calls offering status reports: "We can hear you loud and clear in the Mission!" And complaints: "Answer the fucking phone!" There are words of encouragement: "Rock on." My favorite call: "I listen every night."

One listener calls every Monday during Brad's show, but it's hard to tell if he's a fan or a heckler. "I like Brad," says the caller, who IDs himself as Frosty Freeze and speaks in falsetto. "He's really, really good. Play 'Cherry Bomb.'" Once Frosty realizes our phone number is only voice mail, he starts laying it on thick. "Brad's so sexy. I'm masturbating," he screams before hanging up. Or "You're so dirty. I can smell your socks. Play Ted Nugent."

Some of the callers are friends of the DJs calling to show their support, but most of them have no connection to the station other than listening to it. What I'd hoped would happen actually is: People are finding KPBJ on their own. I'm amazed, since I'm not doing anything to let people know we're on the air. I'm too scared to put up flyers.

My only advertising is the KPBJ T-shirts I've given all the DJs. I don't think any of them wears them very much, at least I haven't seen any of them wearing them, though Abby claims hers was stolen out of the dryer at a Laundromat on Haight Street and that a homeless person is now using it. Not exactly the kind of advertising I'm looking for.

There's a message on the voice mail from Mark Mauer, who works at Bong Load Records in L.A. Bong Load released Beck's first single, "Loser," and is one of the handful of record labels sending me music. Mark wants to know if I'd like the band Lutefisk to stop by KPBJ when they're in San Francisco next weekend for a show. I'm flattered, even if the only reason Mark is trying to get the group on KPBJ is because he couldn't get them on any of the other stations in town. I don't know much about the band, but I like a couple of the songs on the CD Mark sent me, so I say yes.

Lutefisk is the first band to visit KPBJ, and I'm nervous. Not only is it the first time I've allowed anyone into the station whom I don't personally know, it's also the first time I actually have to be Paige in the flesh.

I am on the air, playing the Lutefisk CD in anticipation of the group's arrival, when I hear the gate to the backyard unlatch and the crunch of feet on lava rock. I turn around. A clump of smiling boys are clogging the doorway. I invite them in. The guy with the gravity-defying Jew-fro introduces himself as Quazar. The tall and skinny boy who's missing a chin is Frosting, and the guy who's dressed as a sort of Weezer geek is Watson. Dallas Don, the band's lead singer, isn't there, which is just as well since there's barely enough space for all of us to fit in the "studio."

I'm pressed up against the bathroom door at one end of the DJ table. Frosting is doing his best to avoid stepping on my futon. With everyone clustered together, I pull down the fader on the mixing board and announce that Lutefisk is in the house—literally. The studio isn't at all equipped for guests. There's only one microphone, so I'm forced to do the butter churn, asking questions, then passing the microphone in front of each one of them as they answer. We chat about their record and play a few tracks. I invite them to pick out whatever other CDs they'd like to hear from my extremely limited collection. Quazar picks Ween's "Freedom of '76." Frosting wants to hear Rollerskate Skinny.

Watson attempts to get my attention. "Paige?" he asks.

I don't respond, even though he's standing right next to me.

"Paige?" he asks again.

Still no answer.

"Paige. Paige. Paige. Paige. Paige. Paige. Paige," he repeats, until I realize he's talking to me.

Finally I turn my head, looking at him blankly like a Stepford wife. "Yes?"

The boys in the band exchange glances.

"Can I get through to the bathroom?" Watson asks.

"Sure thing." I press up against the DJ table, and he slips behind me. I am so busted, but I just smile and act like nothing's wrong, as if I were simply temporarily deaf. The band sticks around spinning records for another twenty minutes or so, then leaves for the Paradise Lounge. Later, I notice they've left behind a matchbook. Written around the remaining matches of its interior is a note: "Thanks, 'Paige.'"

About two months after we've been on air, I get a call from a guy saying he works "in the business" and is a "friendly" listener. The reason KBPJ's frequency isn't licensed in San Francisco is because it's a test signal for digital audio broadcasts, he tells me. The next test is coming up in a month. He suggests I vacate my spot on the dial. He doesn't leave his name or phone number.

Two days later, I get another call from the same guy. He says he and a friend tracked KPBJ's signal and are sitting outside my house. I think he's joking, until he leaves another message saying he saw a girl with long red hair and a backpack leave my apartment and walk down the street. He was describing Gabrielle, a new DJ, who had just left my house. I am terrified and turn off the transmitter immediately. He calls again. This time, he's laughing. "Don't worry," he says. "I won't turn you in."

I guess radio triangulation is sport for some tech types.

It's late, well after midnight, when Will comes home with two friends. I don't know either of them, and I don't think Will does either. At least, he's never mentioned them to me before. One is a friend of the bass player in Will's band. The other guy is from Ireland, in town to play a show with Shane McGowan and the Popes. Both of them are wasted.

When Will goes to the refrigerator, I know he isn't looking for anything more to drink. He pulls out the little tin where he keeps his heroin and begins prepping a fix. This seems to be happening more and more since Will and I got back together, but it turns out the junk isn't for him this time. The friend of the bass player has never tried heroin before and is curious, but doesn't want to shoot up. Will "spikes" him, squirting some of the liquid out of the syringe and into the guy's nose so he can snort it. He passes out in less than a minute and doesn't wake up. While Will and the other guy debate calling for help, I am on the phone in an instant, frantically dialing 911. An ambulance arrives within minutes and hauls him away. At least Will has the foresight to drag the guy out to the sidewalk before the ambulance gets here so the overdose won't prompt a police investigation of my house.

A couple of days later, I notice a bunch of CDs are missing. So is Will's stash of tar—about a thousand dollars' worth. Things are getting way out of hand. I'm running an illegal radio station, and Will is not only using but dealing hardcore drugs from the same place. If the cops ever come here, I will be going to jail for a long, long time. Yes, I'm attracted to danger, and true, love is blind, but even I can see this can't continue.

Two years earlier, when Will and I first began dating, he was just beginning to experiment with heroin, doing it once or twice a week. He never kept it a secret from me and didn't seem to mind my hovering over him as he cooked the tar and shot it. The whole process fascinated me, even if I had no intention of using the stuff myself. Watching him shoot up was the ultimate in voyeuristic complicity—the vicarious liberation of my wild side. Being the lover of a user was an inadvertent opportunity for me to experience life's seamier aspects without actually participating—to be bad by association.

These days, I'm not so interested in being associated with his badness. What started as a dalliance is now a three-time-a-day, $200-a-week habit—one he supports with the $400-a-month allowance his dad gives him for no other reason than being his only son, and the little he makes as a small-scale dealer for friends. He doesn't have a job. I pay the rent and buy groceries. I'm resentful, and I'm becoming suspicious of his motives for getting back together with me.

In the time Will and I were apart, I managed to find some direction, but Will's life completely fell apart. When he moved back in, I had no idea exactly how far down he had gone. At some point, the heroin he had considered his musical muse had become an addiction and his enemy. His band split up, and he's no longer writing, just shooting up as often as possible. At least once a night, when the DJs are over, he's getting high in the bathroom. It's depressing to be around him, and I'm tired of playing sugar mama, nursemaid, and psychotherapist. I'm through with being a pathetic '90s version of the Tammy Wynette classic, standing by my man.

When I get a job offer in L.A., it doesn't take me long to accept. Earlier in the summer, I'd freelanced a couple of articles for *UHF*—a style, culture, and music magazine based in L.A. Now its publisher, Scott Becker, wants me to be its editor. Taking the job seems like the best possible way to get myself out of what is becoming an increasingly dangerous situation since, despite my disgust with Will's lifestyle, I lack the heart to kick him out.

As for the radio station, I plan to bring it with me, but I'm giving it a new name—KBLT. I kind of like the sandwich thing, even though Will's warned me: "People are gonna think you're Mama Cass."

CHAPTER 3

Monday, I'm supposed to report for my first day of work in L.A., but it's noon on Sunday and I'm still sipping coffee in San Francisco. Will is pacing the floor in his robe and my slippers, smoking and watching me stuff a plastic Safeway grocery bag with whatever will fit in it—blue jeans, thong underwear, the Godzilla T-shirt I literally bought off some guy's back in Tokyo. I'm taking as little as I can for now. Over the next few weekends, I'll move everything else with a series of rental cars.

Considering that I'm about to leave everything and everyone I know for a place I've visited only twice and hated both times, I'm pretty calm. Still, I'm stalling. I always thought I'd live the rest of my life here.

I love San Francisco. I love catching air as I speed down Gough Street when the lights glow green. I love eating burritos the size of watermelons and drinking the eye-crossingly strong espressos available on almost every corner. I love seeing leather boys hold hands in the Castro and lipstick lesbians make out in the street. I love listening to the live organist play before a double bill at the Castro Theater and hearing hippies drum bongos in the Haight. I love watching fog envelop the Golden Gate Bridge and the sun retreat over Baker Beach. I love hanging out with friends who shoot BB guns in their apartments and dance to Johnny Cash songs at biker bars. There is no place on earth more crazy, grand, lively, inspiring, experimental, inventive, free, and beautiful than this city.

But there is also no place more heartbreaking, demoralizing, lonely, cold, wet, oppressive, and devoid of opportunity. I really do need to get out of here. I need to get away from all the thrift-store hipsters who talk about saving the world but do nothing except shots of Jagermeister—the lazy geniuses who expect to be discovered while twirling their dreads at the five-and-dime.

Will walks me to my bike, parked on a flat road perpendicular to the ski slope of a street I live on and watches me bungee cord my Safeway luggage to the passenger seat. I'm leaving, so of course the weather is beautiful. The sun is beaming, the air crisp as a Michigan apple, as if to ask, "Are you sure you want to leave?"

"I'm gonna miss you, sweetie," Will says, his voice hoarse from junk, his eyes in a narcotic glaze. He extends a hand toward my cheek. When his sleeve slips, I see track marks.

"You'll do great," he says, kissing me not so much passionately as slowly on the lips.

"I love you," I tell him. I should be bursting into tears right now, but I just don't feel like it. He's part of the reason I need to go.

I rev up the bike, slipping on my gloves and helmet while it putt-putts to life. I check the bungee cord to make sure the bag won't slip into my drive chain, get stuck, and buck me off midtrip. I climb on. Will pulls down the chin of my full-face helmet to kiss me through the visor hole.

"Be careful," he says, petting my helmet as if it were hair. "Call me."

As I pull away, I congratulate myself on our drama-free goodbye. Then again, we'll be seeing each other in less than a week when Will visits L.A.

This may be the last time I'll be careening down Sixteenth Street on a motorcycle. It's the first time I'll be traveling four hundred miles on two wheels. Will thinks I'm crazy to do this, even though he's a beneficiary of my foolish behavior. If I'd rented a truck, he wouldn't get to stay in a furnished apartment free of charge for the rest of the month.

I head over the Bay Bridge, battling gusts so strong my bike is leaning even though I'm riding straight. I've traveled this path many times on my way to KALX, but today, as I approach the end of the bridge, I turn in the opposite direction, south toward Los Angeles.

As I near Coalinga, my shoulder blades feel as if they've been split with an ax. By the time I reach Gorman, my hands are numb from vibration. Closing in on Santa Clarita, my ass feels wooden and I'm almost deaf from the buzz of my straining engine and the whistling wind. Worst of all, I'm bored. There's nothing to look at but fallow fields and the occasional cow. This is a stultifying trip by car, even with a boom box and friends. Alone on a motorcycle, it's as much fun as drinking a bucket of spit.

It's a relief to get to L.A. or, I should say, the Valley, just northwest of the city. My friend Ian, who had recently migrated to the Bay Area from the Southland, arranged for me to stay at his parents' carriage house in Sherman Oaks. The door is unlocked when I park my bike and stumble bowlegged into my temporary room at nine o'clock. It's already dark. I fall asleep with the buzz of the road rattling around in my head, and wake up the next morning to the whoosh of Valley traffic headed toward the city.

Show time. My new office is in Santa Monica, the oceanside city that L.A. dead-ends into on its way toward the Pacific. *UHF*, the little fashion and culture magazine I'm about to start working for, and its more established sister publication, *Option*, are in a brick building that is also home to Frank Gehry's architectural studio, a handful of clothing companies, and a smattering of design studios. Inside, it is white. Very white. Colgate white. And modern. The main corridor is at an oblique angle, creating the illusion that the room is a parallelogram. Few of the walls reach all the way to the ceiling.

I don't like it here.

"Can I help you?" a woman asks when I walk in. I can only see her head, which appears to be floating over a countertop.

"Yes, hi. I'm supposed to start work here today. My name's Sue Carpenter. Is Scott here?" I ask, fidgeting with the zipper on my motorcycle jacket.

A voice inquires from somewhere I can't see, "Susan?"

Scott, my new boss, emerges, smiling, from his office. I've met him once before but am struck by his height, or, rather, his lack thereof. He is short, almost elfin, with close-cropped but fuzzy rust-colored hair and a squat, square head. He reminds me of a Rock 'em Sock 'em Robot doll.

"It's good to have you here," Scott says, extending his hand for a shake. "How was your trip?"

Grueling, I tell him.

"Don't tell me you rode down?"

I confirm his suspicions but can't tell if he's impressed or freaked out. He takes me on a tour, introducing me to Wendy, the woman at the front counter, and Mina, the obese calico cat lumbering toward me in the hallway.

For a music magazine, the space looks and feels an awful lot like an art gallery. There are no fawning rock-band stickers declaring fan allegiance plastered to the walls. No trashy rock swag cluttering the shelves. Just tastefully framed photos of Exene Cervenka, Rob Zombie, and other alt-rock demigods neatly lining the angled hall.

Scott introduces me to Mark, *Option*'s shaggy-haired editor. Disinterested, Mark barely looks up from his desk to say hi, as if I were a temp, not a permanent employee. Sandy, *Option*'s assistant editor, is only slightly more friendly. At least her eyes dash to meet mine before diverting in another direction.

I hardly feel welcome. I chalk up their indifference to the Los Angeles/San Francisco rivalry, not knowing that's a one-sided duel. While San Franciscans harbor an undisguised hatred for plasticene, water-hogging Angelenos, few Southern Californians waste any brain time thinking about Freaktown, U.S.A.

Scott shows me to my . . . desk. I'm an editor, but I don't get an office. Those are occupied by Scott, Mark, Sandy, and the ad reps. That puts me smack-dab in the center of the place, where the collective sounds of everyone else's offices ooze over the walls and blend together in an unpleasant sonic soup of whatever music they're sampling or conversations they're having that day. To make matters worse, my computer is communal—the only one with Internet access. I'm bumped whenever anyone wants to check e-mail or cruise around on-line.

This is a good thing, I remind myself. Sure, I had my own office and was making twelve thousand dollars more per year working as a secretary in San Francisco, but I'm working full time and getting paid salary to write and edit here in L.A. I'll adjust.

Getting used to living in a city I'd scorned for years is more difficult. Not knowing anyone or having a permanent place to live, I do what anyone else in my situation would do: I cry. A lot.

Every night after work I ride around weeping while I take in the ugliness that is L.A. Thanks to California's helmet law, at least no one can see me.

I feel completely out of place in this city. Los Angeles is hideous, with its endless strip malls and stucco, perma-haze, and freeways. That's to say nothing of the people. All I can see are the stereotypes. Men who could bench press their pickup trucks. Women whose bodies are so skinny they could slide through doors without opening them, assuming their breast implants didn't get in the way.

I've never been happier to see Will than when he comes to visit at the tail end of my first week in town, but I'm also nervous. I doubt Ian's elderly parents would approve of my having guests, especially a male guest with an affection for needles. In the week I've been here, I have seen Ian's mother only once. She was lounging by the pool when I got home from work one night, and I went over to say hello and thanks. That's the extent of our contact.

The beauty of the carriage house where I'm staying is that its entrance is not visible from the main house, so all weekend I'm sneaking Will in and out, prairie dogging from the door each time to make sure the coast is clear. That plan works fine until Monday morning, when Will's supposed to go back to San Francisco and I head to work. As I'm craning my neck for a parental sighting, I spot Ian's mom reading the paper by the pool. Damn. She sees me. She smiles. I match her tooth for tooth, hold up a finger, and step back inside for a split second.

"Ian's mom's out there," I whisper to Will. "Wait for me here. Don't come out until I come back in to get you. I just need to say hi for a sec."

He seems to understand, but I'm only by the pool for a couple of minutes when I see Will walking toward us.

"Sweetie, are you ready to go?" he asks.

As the smile of Ian's mom flattens into a line, I want to tackle Will into the water. I feel leaden. I suspect I'm out of a place to stay. I take Will to the airport, then ride to work. I've barely walked through the door when I get a call from Ian.

"What was Will doing at my parents' house?" he asks when his call transfers through to my phone. His voice is at the edge of control. There's no greeting. Not even a recognition of my name, as if I've somehow become subhuman. "My parents are old. They are in their seventies. They don't need to be finding bloody cotton balls and needles in the garbage."

I'm voiceless.

"You need to get your stuff out of there."

"Oh my god, Ian. I'm so sorry," I say. I feel awful, but I'm also wondering if his parents are really so bored and/or nosy that they need to rummage through guests' garbage before twist-tying and throwing it away. But they did just see Will, and I suspect I wasn't supposed to have anyone staying with me—definitely not anyone with an affinity for tar and a killer syringe collection. Having been around heroin so much in the past couple of years, I'm almost immune to its rough-and-tumble image. I sometimes forget how seedy and scary it really is.

"I was doing you a favor," Ian hisses. "I told you not to disturb them, and you did it anyway. I can't believe you're doing this to me! You need to get your stuff out of there now. I mean it!"

"Okay. All right. I'll be out of there tonight." I'm having this conversation at my desk, in the middle of the office, and am trying to keep my voice down.

"No. You're not listening! Not tonight. I said now!"

"But, Ian, I can't just up and leave. I just started working here. I can't leave until five."

It's 10:30 A.M.

"Say goodbye to your stuff, then, 'cuz they're going to lock you out! You've got an hour. Do you hear me? Get out!"

He slams down the phone. I'm shocked but shouldn't be. This is a guy who was expelled from high school for beating up his twin brother.

An hour later, Ian calls to badger me again. I'm assaulted one more time before I'm able to leave. I cut out of work precisely at five and speed up the 405. The key to the carriage house is in the deadbolt. I pack my stuff as fast as possible, get back on my bike, and leave. I have no idea where I'm going.

I spend the next several nights renting successively cheaper rooms, bottoming out at a hooker hotel off Sunset Boulevard for twenty dollars a night. I want to kill Will, or at least snip off his dreadlocks while he's sleeping next time I see him. Here I am evicted, and he gets to stay in my place up north for free.

I like to think I can handle anything, but the cardboard walls and bloodstained sheets in my latest rental room are too much even for me. I really do need to find an apartment. I'm embarrassed to tell the people I work with that I've been evicted from the Valley old folks' home. But I am blasting through the two thousand dollars I brought with me, I'm in jeopardy of not having enough cash for a security deposit when I do find an apartment, and the magazine, which pays monthly, won't be doling out paychecks for another couple of weeks. Desperate, I break down and tell Jen, a graphic designer with a big heart and an even larger wardrobe, about the roach hotel I've been staying at for the past couple of nights. She is on the phone immediately, calling up a friend in Hollywood to ask if I can stay at his apartment until I find my own place. He agrees. I am curled up in his comforter later the same night.

Before I moved here, I envisioned running the radio station out of an apartment in Venice. It would be close to work—and the beach. But several people told me that Silver Lake, just east of Hollywood, might be a better match. It's the up-and-coming neighborhood, where all the artists and musicians live, they said. Beck has a house there. So do Zack from Rage Against the Machine, Madonna, at least one Red Hot Chili Pepper, and countless other bands whose names will never be known outside the immediate area.

Sitting at my desk sipping weak office coffee, I flip through the *Los Angeles Times* classifieds. Even if aesthetics isn't an issue, geography is. I need to find a space that works both as my apartment and as a radio station. Personally, my needs are simple. I am looking for a carpet-free one bedroom with as few shared walls as possible. But the radio station has more specific criteria. It needs to be on a hill. It needs to be in a building where the landlord does not live on the premises or, even better, anywhere in the vicinity. It can't be in an apartment building— too many potentially prying ears and eyes. It needs to be large enough for me to escape the music when necessary. And it needs to cost no more than $600. That's all I can afford since I am now making just $26,000 a year.

I come across a listing: "SLVRLK $675 lg 1BD, hrdwd flrs, hi ceil." I call the number and arrange to see it over my lunch hour. Rather, my lunch hour and a half. I'm still getting familiar with L.A. sprawl. It takes me two freeways and thirty minutes to get there.

Exiting off the 101 and heading north on Silver Lake Boulevard, I immediately feel at home. With its rolling hills, the neighborhood seems a lot like San Francisco. Riding along

Sunset Boulevard, I pass a Cuban coffee shop, 99-, 98-, and 97-cent stores, a Laundromat, and a trashy diner. I like the neighborhood's rundown vibe. It reminds me of the Mission in SF. I can't get a read on exactly who lives here, though. There is no street culture to speak of anywhere in this city, and Silver Lake is no different. For all its supposed hip factor, I don't see a single hipster. They're probably all indoors, nursing new tats and preserving their pale skin.

The apartment I've arranged to look at is part of a four-plex—the lower right-hand unit of a hideous, slate-gray stucco number. I know I'm spoiled by San Francisco and its postcard-perfect Victorians, but this place is truly an eyesore, at least from the outside. Nick, a moderately attractive fortysomething guy in boat shoes, is here to show me around. He unlocks the front door, leading me into a jaw-droppingly gorgeous living room with hardwood floors and nine-foot ceilings. The ad did not lie. Oohing and aahing, I trail him through the dining room, then the kitchen, where an enormous O'Keeffe & Merritt stove is parked against the wall, and an alcove that looks out at the Hollywood sign in the distance. Nick takes a U-turn and doubles back to show me the other side of the apartment. We walk through the bedroom, then down a long hallway that is anchored by an enormous walk-in closet. It is soundproofed.

"The guy who lived here was in a band and used to practice drums in there, I think," Nick tells me.

This place must be mine.

I ask who else lives in the building. There are families upstairs on both sides—each with two parents and a single child. Next door: a clothing designer. In the carriage house: a USC law professor.

"Is there carpet upstairs?" If I'm going to run the radio station here, I want to make sure that as little sound as possible can escape the apartment.

There is, he tells me.

"How about the walls? Is it very easy to hear stuff?"

"I haven't had any complaints," he says.

I ask for an application. "If everything checks out, how soon could I move in?"

"Whenever you like."

All the right answers. Nick is my new favorite person on the planet, especially after I ask him where I should send the application. There's a reason for the boat shoes. He lives on a schooner in Santa Monica. I'll have to mail the application to his P.O. box. He will never be here.

This is too perfect. Well, almost perfect. The apartment isn't on a hill but a slope—a mound by San Francisco standards. And I'll be sharing both a ceiling and one entire wall with neighbors. It's farther away than I'd like. My daily commute will be twenty-five miles each way, an acceptable distance by Angeleno standards but a bit of a haul for someone who just moved from a city as compact as San Francisco. It also costs a little more than I'd like, but I'm not going to go the mat with myself over an extra seventy-five dollars a month.

I head back to work, passing a Radio Shack just one block away from my—and KBLT's—new home. Clearly, this is meant to be.

CHAPTER 4

For the next few weekends, I take advantage of the Enterprise Rent-a-Car a few blocks from my new apartment and rent a series of no-frills econo cars. I don't care what the thing is or looks like. If it's got four wheels, a trunk, and unlimited miles for under thirty dollars a day, it's going to San Francisco. My journeys are dual purpose: I can forget about my so-far miserable L.A. life by drinking myself silly with Will and Galen for a couple days, then cram whatever's left of my apartment into the car and cry myself home. It doesn't take too many trips before everything I own is piecemealed to the south—my mismatched and chipped dishes, the same five outfits I always wear, enough secondhand books to necessitate a tire pressure adjustment. My only furniture is an overstuffed suede chair and a futon that is cultivating mold on its underbelly. I take the chair but donate the bed to Will.

During one of my whirlwind weekend trips, I meet with Larry to pick up a new transmitter and sell the Dunifer equipment, which was an unruly child requiring constant attention and discipline. Sometimes the little monster's signal hummed so loudly the music had a permanent bass line of buzz. On other occasions it would get so hot it could no longer stand itself and click off without warning. Fixing its myriad and recurring problems required Gandhi-like patience and more than a little black magic. Even so, the results were short-lived. Larry was making weekly pit stops at my house to tune the beast into submission. Dunifer, while happy to relieve aspiring low-power broadcasters of their cash, was often unavailable when it came to fielding service calls. Whenever Larry calls Dunifer to ask for advice on how to improve our transmitter's performance, he's almost always passed off to an underling who understands even less about the equipment than Larry does.

I should do everyone a favor and dump the Dunifer transmitter where it belongs—deep in a filmy lagoon—but instead I sell it. It works, after all. It just doesn't work well. In a miracle of timing, Chris A. has just called Larry to ask if he knows anyone with a transmitter to sell. Chris is in touch with a Mexican Zapatista who wants one, and Dunifer is out of prebuilt rigs. A sort of goodwill ambassador to the Bay Area pirate scene, Chris is closely involved with all the micro radio operators in town, including the militantly political Latino station, Radio Libre.

Chris passes the ball to Larry, who brokers a late-night deal at the repair shop where he fusses with computer innards by day. It's a nondescript space with brick walls, fluorescent lighting, and mounds of next-to-death machines waiting to be revived. Tonight the only fluorescent on is the one above Larry's workstation. He is giving the transmitter a final tune-up before handing it over.

There's a knock on the steel door. A peach-fuzzed and wide-eyed Zapatista, going by the unlikely name of Dr. Locca, is standing in the entryway. We exchange brief and nervous pleasantries.

"Two hundred dollars?" he asks, raising bushy eyebrows. Dr. Locca doesn't speak English very well but he does know green.

I nod, and he hands over the cash. Larry waves Dr. Locca toward his workbench, where the transmitter is still open for surgery. He points to a couple of metal screws and explains how to tune them. Poor Dr. Locca. Tech talk is hard enough to understand in English. I question whether the good doctor will ever be able to get this thing to work in the remote hills of Mexico while under siege. I feel a pang of guilt. Suddenly I'm feeling like a drug dealer who's cut the cocaine with baking soda.

Larry and I leave for L.A. the next day. Larry has been toiling away after-hours to build the station's new transmitter. It's a hot-rodded ham radio, modified to produce 40 watts—double the power of the one we just sold. He plans to take the week off from work to set it up. We agree to share driving duties for what I hope will be a six-hour trip. I volunteer for the first leg, but Larry is clearly unnerved by my 20 percent addition to the speed limit and unrepentant tailgating. I can almost hear the swish of his corduroys against the cloth of the passenger seat as he shifts in discomfort. About an hour in, we pull over for gas. When I return from paying the attendant, he's at the wheel. For the next hundred miles it's his turn to watch me claw the hand grips as he drives exactly the speed limit—no faster, no slower. After my bladder demands a stop, it's a free-for-all. Whoever drives depends on who gets to the steering wheel first after we've pulled off the road.

As I feared, it is practically Cinderella's bedtime by the time we get to L.A., but we don't go to my house. There's work to do. Without unpacking, we head for the hills of Silver Lake, where a slow fog drifts over the car as we roller coaster through the patched streets to find the highest hill in the area. The radio equipment in the trunk rattles, then clunks, G-forced to the back as we strain for the flat land at the peak of Micheltorena, about a dozen blocks northeast of my house. The lights of Hollywood are twinkling in the distance as we pull to a stop in front of an abandoned convent, killing the engine and head-lights but leaving the ignition on so we can listen to the radio.

There is something uncomfortable about the two of us sit-ting here in a parked car in the dark this late at night. I just hope this isn't a well-patrolled area. It would be difficult to con-vince a police officer we're not here to trade favors but to figure out what frequency to use for a pirate radio station.

There's better reception up here than at my house, which is, unfortunately, at a significantly lower altitude. Up here, we'll have a better idea of where we can broadcast without stomping on anyone else's air. In the Bay Area, all the available frequen-cies have been identified and exploited by the movement's pio-neering pirates, but here in L.A., Larry and I have to figure it out ourselves. Rather, Larry is figuring it out. I'm just watching. I'm still coming up to speed on this radio thing.

We begin our dance across the FM dial, starting at the lower left end, tuning past the college stations broadcasting out of Pasadena City College, Loyola Marymount, and Santa Monica College. Crossing FM's Mason-Dixon line into commercial ter-ritory, we hear the heavy bass of hip-hop on 92.3, noodling gui-tars on classic rock Arrow 93.1, and oompah-pah Mexican ranchero at 94.3. Larry pauses middial, spinning the knob left toward the rock station on 95.5 and right toward Mexican pop on 96.3 like a safecracker listening for that special click. He pauses on 95.9 and turns up the volume, training his eagle ears on the white noise blaring from the speakers. He digs in his backpack and pulls out two pieces of paper. One is a topographical map of Southern California. The other is a list of L.A. area radio stations—where they broadcast from, on what frequencies, and at what power. All of this factors into where we can squeeze in on the dial. It won't be 95.9 because it's too close to another licensed station.

He continues spinning the knob upward, listening to snippets of oldies on 101.1, club music at 103.1, alternative rock on 106.7, and Mexican pop where the dial tops out at 107.1. He's not listening to the music so much as the snow on the frequencies in between.

"Oh jeez," he mutters.

Larry, who reads technical manuals like I read *Elle,* is balancing the two sheets of paper on his lap. Beads of sweat are beginning to leak from his forehead. Slowly, methodically, he tunes through the midsection of the dial with the knob, reading the LED display and reciting each frequency out loud as he slowly spins by.

"104.3," he says slowly.

"105.1." Sigh.

"105.9." He runs a palm over his moist forehead. "Gosh, Sue. I dunno."

"What's up?"

"I dunno. There's nothing here."

I prompt an explanation with a cocked eyebrow.

"L.A. I didn't even think, but there are so many stations here. There's no place for KBLT. Oh jeez."

Now I'm beginning to sweat. The radio station is with us in the trunk of the car, and now there's nowhere to put it?

"Larry," I moan. I'm grumpy from the long trip and crashing from the day's diet of Coca-Cola, potato chips, and Pop Tarts. I lean my head against the cool window and close my eyes.

Larry glances down at the papers and rests the dial on

104.7. Most of what we hear is static, but there is also a trace of hip-hop.

"I dunno. Maybe it fits here."

I open my eyes and sit up. According to Larry's lap, there is an urban-format station on 104.7 in Oxnard, about sixty miles north. Broadcasting at about 50,000 watts, the station should come through loud and clear, but the music is barely audible. It's mostly fuzz.

"The Hollywood Hills? Santa Monica Mountains?" Larry asks the window. The mountains are blocking Oxnard's signal from traveling all the way into Los Angeles. We drive over the hill a little to make sure the solidity of the signal is consistently weak. It is, and it will be even weaker at my house. It looks like 104.7 is our frequency. We breathe a collective sigh of relief, then rev up the car and head home to my skeletally furnished apartment. I pull the cushions off the secondhand sofa bed I just bought for the living room and dress it with sheets for Larry, then conk out in my own bed without bothering to scrub the road scum from my face or fish my toothbrush from my travel bag.

The next morning, I return the rental car, then head to work on my bike, leaving Larry with the dirty work of installing at least the beginnings of the station. I feel guilty, but I know so little about electronics at this point that my only value to him would be as a human tool belt, holding tuning wands and solder. So, back to the day job.

It turns out that running a magazine is a lot of work, and not only have I never done it before, I have less than a month to pull together the entire January/February issue of *UHF*. I get to the office at about eight-thirty, hoping for an extra half hour of concentration time before the *Option* magazine music critics arrive, and I'm subjected to the sonic sludge of competing stereo systems. Wendy and Mina are the only ones here. Wendy, who commutes thirty miles each way from Pasadena, gets in earlier than me to beat traffic, which, even on good days, doesn't flow so much as budge.

At my desk, a stack of stories threatens to obscure my computer monitor. Scott has assigned most of the copy for this issue, but I am editing it. Among the stories awaiting my red pen are a column on how to discern a fox from a wolf at New York City nightclubs, a profile of some San Francisco clothing designers I've never heard of, and movie reviews by Fenton De-Bell, aka John Lockhart, one of the magazine's ad reps. Scott's also asked me to caption fashion features—a ten-page photo spread on Brooklyn homeboys posing tough on the A train, a sacrilegious shoot with pouty-mouthed altar girls that would have Popes John I, II, and III twisting in their coffins.

I'm also supposed to talk to a bunch of musicians about what they wear, an assignment I dread. Every time I interview a celebrity, both my brain and my funny bone desert me, making me sound more witless and dumb than Dan Quayle. It's hard enough to talk intelligently about their art, let alone something as superficial as style. Then again, this is a fashion magazine.

UHF, unestablished as it is, shouldn't be able to draw the musical star power it does. But Scott is working his *Option* connections to build *UHF*'s street cred, which is in the red at this point. We hope to turn that around with this issue. We'll be running a fifteen-page story called "Dressed for Success," with photos of Moby, Foetus, Cake Like, DJ Spooky, the Meat Puppets, and a handful of lesser groups whose CDs are destined for the resale bin. I've been chasing interviews for the better part of a week, ultimately talking to about half the bands.

Some are more indulgent than others. J. G. Thirlwell, aka Foe-tus—mastermind of monosyllabic industrial classics *Nail, Hole,*

Deaf, and *Ache*—is a surprisingly good sport. Even better, he gives great sound bite. The self-proclaimed tuxedo junkie tells me he aspires "to be a year 2100 Gary Glitter without the paunch and with a bigger bulge in my pants." DJ Spooky, the postmodern, postrational DJ whose mixes blur the boundaries between music and art, is also a sweetheart, even if his music bores me. He's one of those rare intellectuals who spare you the dissertation and don't shove their brains down your throat. He's also something of a humanitarian. When I ask where he gets his clothes, he gives new meaning to the term *streetwear.* He buys them from the homeless. Meat Puppet Curt Kirkwood, twin brother of his forever de- and retoxing bandmate Cris, is a bit cagier and condescending with his answer. "We overlook a lot of miracles," he tells me. "Think about Eli Whitney and the cotton gin, or the loom. Everyone should be thinking how clever it is we even got our arms through these little holes." I appreciate the humor, I just wish it weren't at my expense. I'm just doing my job.

Paige Jarrett takes the credit for this story, because Sue Carpenter has already contributed one to this issue—"Odd Jobs: Making Do Before You Make It Big." It looks bad to have too many articles written by the same people. It makes the magazine appear understaffed and cheap, which, though true, is not what you want the readers to think. About a third of what runs in *UHF* is written pseudonymously to perpetrate this scam, including this issue's record-review column and an interview with ooh-la-la French actress Julie Delpy, both of which are written by Scott but attributed to Benjamin Diaz.

No one at the magazine suspects Paige Jarrett is anything more than a pen name I plucked from my nether regions. I haven't told them this fake byline is actually my alter ego. Being an editor is demanding enough work. Simultaneously launching a radio station is like spit-roasting pig for a side dish while cooking lobster for the main course. I have no idea how Scott would react if I told him. Being a music lover, he might think it's cool. But being my boss, he might also think it's too big a distraction and drop-kick my ass back to San Francisco.

I return home to find Larry squatting on the floor in a corner of my bedroom behind the door leading to the hallway. His thin plaid shirt is undone a couple of buttons too far, exposing little curlicues of dewy chest hair. He looks suspect, but I know what he's up to. Scattered around him are speaker wire, antenna cable, black spray paint, a fan. I recognize a battery-looking thing with the red and black connector terminals as a power supply, a smaller black gizmo as a power meter, and the silver metal thingy with screws as the piece that drives the system.

I have no clue how these things work. I just know enough to identify the individual components and what they do—enough to ace a multiple-choice quiz if someone wanted to give me one. But if Larry left and I had to put the pieces together myself, I'd probably blow out the power for my entire block.

The transmitter is in my bedroom—far enough away from the studio equipment that it shouldn't cause distortion. We plan to snake the antenna cable under the door, through the pantry, out the hole we've drilled through the window frame, and up to the antenna, which we intend to mount to the neighbors' back porch while they're out of town for Thanksgiving. The transmitter is, unfortunately, visible from the bedroom window, a room that has nothing in it except a secondhand mattress. There's no place to hide the transmitter. My understanding is that law enforcement officials are legally allowed entry if they see ample grounds for search, which would clearly be the case here since I do not have curtains. I'm not one for window coverings. I hate blinds, shades, drapes—all of it—but I make them a priority now.

On my way home from work the following night, I pick up some semisheer white cloth and a curtain rod from a fabric store. Later in the week, I'll tap my inner Girl Scout and sew them together. I also stop by my new mailbox service to see if any music has come in for the station's imminent relaunch. It has. I open all the padded mailers on the spot so I can maximize the space in my backpack and stuff a couple of dozen CDs into my bag, handing the trash to the man behind the counter, who feigns a smile.

When I get home, I notice that Larry has moved his base of operation to the dining room, a room in name only. There's nothing in it except for Larry and a slew of copper pipes, which he has been cutting with a hacksaw, smoothing their ends with sandpaper. The pipes are for the antenna and need to be specific lengths to work with our frequency at 104.7.

"Where'd you find the pipe?" I ask Larry. "I've ridden around this neighborhood, and I haven't seen a hardware store anywhere around here."

"Orchard on Sunset," he answers, not looking up. He blows the metal dust off the end of a two-foot tube.

"Isn't that a couple of miles away?"

Perched on one knee, Larry looks up sheepishly.

"You didn't take a cab, did you?" Larry makes just as little money as I do, maybe even less.

"I walked. Man is it hot down here."

"Larry!" I scold. "If I'd known you were going to pull this vagabond plumber stunt I would have sprung the extra sixty dollars to keep the rental car an extra couple of days. I can't believe you did that, you freak."

Actually, I could. Larry and I share a common affliction: radio mania. We will do whatever it takes to get the station up and running. Tonight, that means a second, last-minute sprint back to Orchard Supply Hardware to buy the stuff Larry needs to install the antenna tomorrow while I'm at work. We end the night with dinner at the House of Pies a couple of blocks from my house.

I order a BLT. Larry devours a piece of cherry pie for an appetizer, then polishes off a Cobb salad the size of my head. It isn't articulated, but Larry and I have a deal: He donates his time and technical expertise, I pick up the tab for hard materials and food, and together we make our collective vision a reality—something neither one of us could accomplish alone. A quintessential gear head, Larry is far more comfortable around alligator clamps and transistors than he is around people. He has the know-how to build a radio station but is way too shy and private a person to populate it. I'm not much of a people person either, and I don't know an alligator clamp from my ass, but I am organized, I am motivated, and I am resourceful.

Larry finishes building the antenna the next day while I'm at work and has it mounted by sundown. Hearing my bike as it buzzes up the street and pulls into the driveway, he walks onto my back porch. His smile is aw shucks, and his hands are buried in the front pockets of the faded gray corduroys he's been wearing the entire three days he's been in town.

"Well?"

I look up the wooden staircase that runs behind my bedroom window and up to the neighbors' back porch. There, next to a laundry line and a potted palm, is the antenna. It almost looks like a fork, if its prongs were on sideways. I wonder what my neighbors will think when they see this thing. I'd asked them earlier if I could use their porch for my "television" antenna, but this homemade contraption, with its spray-painted black pipes, doesn't look like any television antenna I've ever seen.

"It's lookin' weird, but lookin' good, Larry. Did you have any trouble getting it up there?"

"Nah." He shrugs, even though the rings under his arms say otherwise. "Want to check it out?"

"You mean you've hooked up the transmitter, too? You're a god."

When I left that morning, the studio was set up as my home stereo. Now it's a radio station. I power it up, then put in one of the CDs I picked up from my mailbox service yesterday. It's Ken Nordine's *Colors*. I have no idea who Ken Nordine is, but I listen to everything I get, and now's as good a time as any to preview it.

I turn up the volume in the studio and press Play. A jazzy ode to the color olive comes blaring through the speakers. Larry and I both have cheap portable radios with earplug headphones. We put them on and tune to 104.7. What we're hearing in the studio matches what's coming through on our handhelds. So far, so good. We walk out the front door and hang a left toward McDonald's. We intend to kill two birds with one stone. We will test how far the signal travels and get dinner. By the time we get to Sunset, Nordine has smooth talked his way through the better part of the color spectrum, with tributes to lavender, burgundy, yellow, green, and beige.

Larry and I are quite the sight. No one walks in Los Angeles at night except hookers, and we are not only walking together but silent, acknowledging each other only when we happen to make eye contact, at which time we smile and raise our radios in a sort of toast. The transmitter is working. We celebrate with Big Macs and supersize fries, then turn around and walk home to songs about mauve, fuchsia, sepia, and cerise. KBLT's in business.

For the next month, I'm stricter than the military with my hours. Every night after work I flip on the transmitter at eight o'clock and spin records until ten, going on faith that someone—anyone—will happen upon KBLT, even if the five times I check voice mail every hour tell me otherwise. No one is listening. At least no one calls.

New Year's Eve is more of the same. It's just slightly more depressing to be alone in a closet playing music for no one but the ether when everyone else is out drinking with their friends. I've never liked New Year's—the need to pretend I'm having a good time, that strange sadness I feel when the clock strikes twelve—but this one is shaping up to be my worst. Earlier in the day, my boss's neighbor gave me a cat, who promptly ran away through a hole in my closet. Then, as I'm curled up under my blanket before midnight, I wake up to a cracking sound, followed by the hollow thud of metal pipe connecting with concrete. I slip on my robe and shuffle sleepy-eyed out the back door. By the moon's light, I strain my eyes to see what it is. Part of the antenna is rolling around in the parking lot. It's those goddamn Santa Anas I'm always hearing about—that "persistent, malevolent wind" as Joan Didion says. I've been listening to them whistle around the building for weeks, but I never suspected they could rip my antenna in half, hurling it in pieces to the pavement. This is simply too depressing to deal with now. I just pick up the cracked pipes to stop them from rattling, bring them inside, and jump back in bed, pulling the covers over my head and pretending this is not my life.

Sleep won't come, so I cut my losses and get up early, calling Larry to share the bad news. I beg him to come to L.A. for emergency repairs. If I buy him a plane ticket now he can be here midafternoon, but Larry can't visit for another week. He offers to walk me through fixes over the phone. I don't like the idea. Intensive meddling with electronics is among my greatest fears. Hoping to get out of it, I tell him I'm a woman of words, not science. He doesn't buy it. I play helpless: "But I don't know what I'm doing!" I try flattery: "Have I ever told you that you've really got a way with that soldering iron?" I whimper. I pout. But he holds firm. Unlike KPBJ in San Francisco, where Larry was listening in and came to the rescue, no one's left any calls on the station's voice mail. I'm on my own. I really do have to fix this myself.

I'm allowed to go home early from work the next day to hunt for my lost cat. I "here, kitty kitty" for a block in each direction and put up flyers, but my cute new kitten remains in hiding, so I use the remaining daylight hours to dismantle and repair the antenna. I'm on the phone with Larry for two hours reconfiguring and soldering wires, alternately cradling the phone between my chin and shoulder and resting the receiver on the floor, shouting my comments toward the mouthpiece while I butterfinger the parts. But I fix it.

With just a few remaining rays of sunlight, I'm alone on my neighbors' porch, trying to slide a ten-foot pole into its mount on the vertical beam that supports the balcony. The antenna is top-heavy with the two prongs protruding horizontally from its top. It's difficult to keep it upright. Losing my balance, I accidentally step in my neighbors' hibachi, prompting one of them to emerge from the kitchen and scold, "At this hour? It's dark!"

I try to hurry it up, but the pole is too heavy for me to keep it up straight. I remove a five-foot segment of the pole to make it more manageable. It takes a while, but I finally manage to get it in. I'm proud to have fixed it but bummed that I now have only half the antenna strength and half its height, assuming it even works. I turn on the transmitter. Playing the new Yo La Tengo record, I walk around the neighborhood with my Walkman to test the signal strength. Yeah, there's a signal, but it's horrible. Maybe it's a good thing no one's listening.

CHAPTER

Most L.A. newbies take their acting tests on a studio lot. Mine is at a table outside a South Pasadena coffee shop, trying to pass as the flesh-and-blood Paige. It's inevitable that I will begin to meet the people I've so far known only through the mail. Los Angeles is, after all, the belly of the beast of the record industry, home to hundreds of labels, dozens of which have been feeding me music for the past few months.

I'm meeting with Mark Mauer, the Bong Load Records radio rep who arranged Lutefisk's visit to KPBJ a couple of months ago. Mark was among the first to respond to KPBJ's clarion call for free music. He is also my welcome wagon to L.A.'s indie rock scene—the first person to take pity on the new girl in town and invite her out for a drink.

My first few months in town have been pretty lousy. Yeah, I'm finally working in journalism and my radio station is up and running, but I have no friends. I'm feeling alone, pathetic, and unsure of what I'm doing. I'm thrilled to be getting out of the house on a social call, even though my self-esteem's in pretty bad shape. I'm nervous, awkward, and feeling uncomfortable in my own skin.

Mark and I have spoken on the phone maybe a half-dozen times, mostly to coordinate the Lutefisk interview. All I know about him is that he seems genuine and laughs a lot. He's given me absolutely no reason to mistrust him. I mean, I should be able to tell him my real name, but my San Francisco bias against Angelenos is strong. Sure, Mark's giving me records today, but tomorrow he might decide the station would sound better broadcasting from the moon and call the FCC. It's best I keep my real identity a secret.

Mark and I meet on a Saturday, one of those hazy gray afternoons where it's hard to tell what is natural cloud cover and what is smog. Mark is already there when I wheel up, sitting outside and warming himself with a cup of coffee and a black pea coat that makes his ultrapale skin seem blue. He is a sliver of a guy with a healthy head of black hair and wire-rimmed glasses set before almondine eyes. His face launches into a grin as I approach the table. I immediately like him.

This isn't a business meeting so much as an opportunity for Mark to see what kind of fool would start a pirate radio station in L.A. As far as either of us is aware, mine will be the first. We chitchat—about KPBJ's demise, KBLT's arrival, the reason I moved to town. I tell him I'm an editor but don't say where. He congratulates me on selecting Silver Lake—not Venice—as my home base, and reconfirms the hood's hipster rep, even if my sightings of them have been rare.

"Maybe you're going to the wrong places," he offers. He invites me to see a show later that night at Spaceland, a trendy Silver Lake club where indie rock bands earn their gas money.

We're supposed to meet at ten, but I'm late. I would have

been on time, but the sign—in pink and blue neon—reads DREAMS OF L.A. not SPACELAND. Unaccustomed to the cryptic hipness of L.A., I ride by it a couple of times before figuring out what it is. The place is cheesy—Swiss cheesy. The exterior is a faux Teutonic in stucco. The entryway is mirrored and plastered with flyers for upcoming shows. With its trashy disco interior and hotel conference-room tables, the club has a sort of dumpy charm.

Inside, it's about as full as a church on Wednesday. It's still a little early. The true night crawlers have yet to arrive, but there's at least a smattering of boys in Dickies with wallet chains, Bettie Page ripoffs, and heroin-thin girls wearing baby tees. I spot Mark front and center of the stage—prime heckling distance for the Invisible Men, a group he describes as a surf band fronted by Don Rickles. Mark and his friend Leah recently released a seven-inch of the group on their own Boobie Prize record label. Diehard fans, they travel to every show and relentlessly lob insults from the audience. In defense, the band members pelt whatever they can find on stage into the crowd, invariably missing their targets and making a mess. That's why the Invisible Men have been barred from nearly every club in town, Mark tells me. I see why, once they hit the stage. Wrapped in full-body gauze and wearing Ray-Bans, they can't see.

Tonight, there are no flying cans and bottles, but the show is just as entertaining. The lead singer climbs into the rafters. Hanging from his knees, he continues to sing and play guitar until the instrument drops from his hands, at which point his gauze gets stuck and he engages in a minute-long battle to free himself, finally dropping to the stage on all fours in a heap of humiliated gauzy showmanship.

After the Invisible Men finish their set, I get up from the table to refill my pint of Red Hook and see the silhouette of a familiar Muppet-like mop top, backlit by the bar. I tap the shoulder of the head it belongs to.

"Quazar?" I'd met the Lutefisk drummer a few months ago when his band stopped by KPBJ.

"Hi!" he says, recognizing me but struggling for a name.

"Paige. From KPBJ," I prod.

"The pirate station. Right. Mark told me you were coming to town. How's it going?"

"Okay, I guess. Weird, actually. It's hard to adjust." I don't want to dump, though I could go off about how much I hate the burritos here, about how I detest the city's cement and sprawl, but I don't. There's no faster way to end a friendship before it's begun than to whine.

"Well, I wouldn't know. I've lived my whole life here, but I think it must be hard."

For a guy in a well-respected band in L.A., there's not even a trace of ego in Quazar. In fact, there's a real warmth and innocence to him. His mouth is permanently cocked in a slight smile. Even in the dark, his eyes are twinkling.

"I'm here with Mark. Do you want to say hi?" I ask, not knowing the two are close friends.

We head across the beer-sticky carpet to the table and have a seat. Mark has been joined by two business-looking women whom Quazar already knows. Mark introduces me as Paige, but one of the women doesn't hear him over the music and asks me to tell her again. "Sue," I say loudly, before realizing my mistake. "I mean Paige," I say. She looks at me quizzically, but doesn't dwell on it because she couldn't care less. She turns to Mark and resumes her conversation.

I lean back in my chair, take a drink from my pint and match the glass to its wet ring on the table. My eyes make their way around the room, eventually settling on Quazar.

"Hi, Paige!" He smiles and waves, even though he's sitting only two feet away. "Are you all right?"

"Yeah, I'm good." I'm lying. I always feel uncomfortable around people I'm just getting to know, and Quazar is staring at me. I'm feeling studied.

"So tell me what happened to KPBJ?"

I give him the short and glossy version. "There were a lot of problems, and then I got a job here."

"Yeah, Mark told me. Sorry to hear about that. He said you're going to be starting another station down here?"

"I am, but it's not KPBJ anymore. It's KBLT. Different station, different name."

He laughs. "Just like PBJ? Peanut butter and jelly. Now bacon, lettuce, and tomato."

"I couldn't resist. I was just so pleased with myself for coming up with that one."

"That's cool," he says. He's still laughing. "Is it up yet?"

"Yeah, but I'm the only DJ. I don't know anyone here."

"I could hook you up with some people who'd probably want to help out," he offers.

"You do?" I want to plant a wet one on his cheek.

"Yeah. I know a few people at KXLU, some people at labels who might be able to get you some records. I'll call some of my friends."

Quazar is the Kevin Bacon of Silver Lake. He's connected to pretty much everyone in the scene. Thanks to him, I get more phone calls in two weeks than I've had the entire time I've lived in L.A. Calls from people who want to be DJs. Calls from Dave Sanford—owner of a college radio record promotions company and co-owner of No Life Records; Carolyn Kellogg—West Coast editor for the music 'zine *Fizz*; Kerry Murphy—radio promotions director at Slash Records; Chris Carey—a radio promoter at MCA Records; Matthew Semancik—a college radio promoter at American Records; Bill Smith—radio promotions manager at Rhino Records; Rudy Provencio—a radio promoter at Priority Records.

I'm seeing major red flags. Not only are all of Quazar's friends in the music business, a lot of them work in publicity, which makes me nervous. KBLT is supposed to be radio for the people by the people, not a shill for the industry. Why did I even bother bringing the station with me if it's going to be like every other station bogarting the FM dial?

Don't get me wrong. I'm ecstatic to have Quazar's help. He's given me enough people to keep the station on air for two shifts every night—and all of them seem perfectly nice. But what's to stop them from using their direct access to the airwaves to push their own labels' products? Nothing.

There weren't any rules about what the DJs could or could not play at KPBJ. Sure, I preferred that they steer clear of Cher and Lawrence Welk, but they had complete creative control. I really don't want to be the music police at KBLT, but I don't know if I can sleep at night if I let these people deejay. Okay. So screw me and the liberal white horse I rode in from San Francisco. I'll admit I'm suspicious, judgmental, and too goddamn PC for my own good. Quazar's friends haven't even set foot inside my house, and already I'm accusing them of this fantasy low-rent payola scheme. Fuck it. How else am I going to staff the station? Stand on the street corner with a sign? I start calling Quazar's people back, arranging times to train them on the equipment and assigning shifts. Live and let live.

The next few weeks are an indie rock parade as I train about a dozen DJs—enough for the station to be on the air from eight to twelve each night, just like San Francisco. There are a couple of girls with lunchbox purses and Kool-Aid hair, but the vast majority are boys who seem to have pilfered from their grand-fathers' closets. Overall the look is thrift-store chic; the music, a mix of old-school punk, vintage art rock, and college radio fla-vors of the day—Cibo Matto, Palace, Cornershop.

Coming over to my house, some seem to have packed their entire record collections, carting milk crates, postal cartons, and laundry baskets filled with vinyl for their shows. Others carry their CDs discreetly in backpacks. Many come bearing gifts of music for the station library in exchange for the plea-sure of sitting alone in a closet, spinning records, and talking to imaginary listeners. I have yet to receive a single call on the KBLT voice mail.

The studio equipment is fairly simple. Radio Shack's finest. A chimp could learn how to cue up records on the turntables, slide the fader back and forth on the mixing board, and press Play on the CD players. It takes about fifteen minutes to learn and—presto!—insta-DJ.

It's a sort of budget college-station setup. The only differ-ence is that the DJs need to keep my address a secret, be on the lookout for white vans, and know what to do if the FCC stops by for a quick chat and frisk. Like I said, simple. Walking the DJs through the equipment, I correct their mistakes as they spin records without turning up the volume or play two CDs at the same time. Luckily for them, there are no listeners.

Quazar and Mark Mauer deejay a few times, but everyone else who stops by is a complete stranger to me. Kerry Murphy is among the newcomers. It's a Tuesday night when a pigtailed blonde appears on my doorstep wearing a Daisy Duke gingham shirt and Curious George backpack bulging with CDs and seven-inch vinyl. She is extremely cute. And young. And short, at least from my platform-boot-enhanced perspective. I've got at least a half decade and maybe eight inches on her.

"Paige?" she asks, her voice muffled from talking through the glass door. "I'm Kerry," she says, mouthing the words without sound.

We both laugh. This girl's all right. I unlock the deadbolt.

"This is so *I Spy*." She looks back over her shoulder and steps into my living room.

I invite her to have a seat on the couch just inside the door. She sits, opens her backpack, and extracts a couple of CDs and a fresh pack of Camel Lights.

"These are for you." She hands me a couple of Soul Coughing records from Slash, where she works. "Quazar told me you might need some help getting music for the station."

I do. "Always. Thank you."

I offer her a beer. "It's home brew," I warn, getting up from the sofa. "Hope you like yeast."

I return with two Alzheimers and hand one to Kerry. She looks at the label—a cartoon drawing of a head on a sled—and laughs.

She takes a sip. "I think it's so rad you're doing this," she says. "Aren't you scared? I mean, the FCC could come by here at any time, couldn't they?"

I nod. "There are instructions in the studio on what to do when it happens."

"This is so *Pump Up the Volume*."

We head into the studio and I point to the Xerox tacked to the wall: WHAT TO DO WHEN THE FCC KNOCKS. Most of the information is from a pamphlet put together by the Committee on Democratic Communications. It's a simple list:

1. Before answering the door, turn everything off and unplug the transmitter. The FCC is legally entitled to take

not only the transmitter but everything attached to it, i.e., the studio equipment.

2. Do not open the door. Talk through the glass.

3. Do not answer any questions pertaining to the radio station. Just like the movies, anything you say can and will be used against you in a court of law.

4. Do not let them in unless they have a search warrant, in which case you are legally obligated to do so.

She sets her beer down on the DJ table and surveys the closet. The room is decorated with a single Boss Hog poster and promo stickers for the S.F. Seals, Whale, and any other band whose record I've received in the mail lately. There are about a hundred CDs in racks on the wall, maybe twenty records.

"How can I have more records than you, but you're the one with a radio station?" Kerry asks. "Quazar was right. You really do need some music around here."

She pokes her head into the living room, then turns back into the studio. "Yep. That's what I thought. You've got more books than records. How did that happen?"

"I know. It's kind of embarrassing to run a radio station without any music, huh? I wasn't a music collector when I started this thing in San Francisco."

"I guess not," she says, checking to see what I've got. "So, what were you?"

I'm tempted to tell her I was a person who loved music but knew nothing about it, a regular girl on a fast track to nowhere, but I keep quiet. It's time to get on the air. Kerry has volunteered to deejay Fridays at 8 P.M. That makes her show the first of the night, which means she's responsible for turning on the transmitter. It's in the bedroom, I tell her, and lead her down the hall. I feel like a guy who's just used the worst pickup line on the planet: "Hey, baby, want to see my transmitter?" I'm feeling a little weird, but Kerry doesn't seem at all freaked. I flip on the light.

"That silver box there in the corner, that's the transmitter. The black thing next to it is what powers it. You'll need to turn on both of those and the fan, otherwise it'll overheat. Got it?"

"So as soon as I turn that stuff on, whatever was happening on 104.7 suddenly turns off for everyone, and it becomes my show?"

"Simple but true."

We go back to the studio, and I point out the power strip underneath the table. She flips it on.

"And now you're ready to deejay," I tell her. "Grab your records and have a seat."

She collects her backpack from the living room and sits down in the three-dollar studio chair I bought from a thrift shop around the corner. The top cushion is held to its frame with nails and squeaks when she leans back. She laughs.

"Pirate," we say in unison.

She unloads CDs by Stereolab, Archers of Loaf, the Halo Benders, and vinyl by a bunch of local bands I've never heard of—Pop Defect, Longstocking. The only Slash records in her bag she gave to me. My worries that industry people will only play their label's bands are unfounded, at least with Kerry.

I walk her through the controls, since she's never deejayed before, and we spend the next couple hours smoking, chatting, and drinking while she spins. For a newbie, she does pretty well. She only forgets to turn the microphone off once. Kerry leaves at ten, saying she'll be back on Friday for her regular show.

It's a relief to finally have some people to split DJ duty. But it's also difficult to have strangers over every night of the week while I cook dinner, talk on the phone, clean the house, get ready for bed, and go to sleep. I am feeling self-conscious in my own home. It's like trying to adjust to a new roommate, only the roommate keeps changing.

As awkward and anxiety-ridden as it is for me, it is doubly so for the DJs who are coming to an unfamiliar apartment inhabited by a stranger to play music on a radio station no one has ever heard of and that could get shut down at any moment. When I train Jasmin—a nineteen-year-old counter girl at East Side Records, a used-record shop not too far from my house— she is visibly unnerved when I take a shower, then roam the house in my robe and slippers. Every time I walk into the studio to ask if she has questions, the look on her face says "Go away!" I'm unwelcome in my own apartment, and I'm the only one who lives here. What a mind-fuck.

For the most part, the DJs are on their best behavior. They don't know me very well, so they're respectful of my space. If they want to smoke, they ask permission. If they eat or drink, they clean up after themselves. If they bring friends, they call ahead for approval. A couple of the DJs are so protective that I sometimes feel like a mafia don. One girl blindfolds her friend before bringing him to the station. Another parks a block away and makes his guests walk through my neighbors' backyards to keep my address a secret.

There is low-level paranoia among the DJs—a sense that at any time they could show up for their shift and KBLT would be gone, my front door swinging in the wind, the studio space cleared of everything but speaker wire. So when they see something suspicious, they let me know. If a DJ tells me there's a weird old coot picking his teeth in a pickup truck just outside my front door, I check it out. Personally, I'm on the lookout for windowless white vans, even though no one has ever told me that's what the FCC people drive. I don't know where I got that idea, but now that it's solidified in my mind, there is almost always a white van parked somewhere on my street. I spend hours sitting on the cement steps in front of my apartment feigning prolonged interest in my shoes while I investigate these false alarms, my investigations consisting of little more than occasional glances at my suspects until they drive away.

I'm gradually getting accustomed to answering to the name Paige. I've flubbed only twice, most recently when I called one of the DJs and he answered his phone "Who is it?" instead of "Hello" and I told him "Sue." Otherwise few, if any, of the DJs even suspect Paige is a fake name.

I'm friendly but distant with most of the people coming into the station. If they ask too many personal questions, I become evasive. I have my identity to protect. But I'm also uneasy. Never before have I felt so surrounded by people and yet so alone, but that is my fault. I've always been a loner, usually with only a handful of close friends. To suddenly be at the epicenter of hip is uncomfortable. In my mind I am still the same shy, awkward, nerdy girl I was at age five, even if the motorcycle and pirate radio station give a different impression to the DJs who know almost nothing about me.

They have few clues as to who I am. The only beings they've seen me with are Larry and my two cats, Ralph and Mitzi— who reappeared in my living room two weeks after she'd run away. My house is sparsely decorated, which is partially a function of my artsy-fartsy minimalist aesthetic but mostly because I have almost no money. Whatever cash I do have I funnel into

KBLT, not furniture. The only object in my dining room is an imitation tree trunk that appears to be growing from under the floor and up through the ceiling. My living room has only a couch, a chair, a bookshelf, and two pieces of art—an Elmo doll I've crucified on a wooden cross and a greeting card display mounted on two plastic racks above the sofa. The cards were a friend's MFA project. They are real Hallmark cards with altered text, but I doubt any of the DJs have picked one up to know. I suspect some of them think they are on display to prove I have friends, kind of like those people who keep the photographs that come with the picture frames they buy.

As the station settles into a schedule with regular nightly hours, KBLT is finally getting calls. They're from the DJs, who now have a vested interest in cultivating a listenership and are concerned with the signal's reach. There are more or less nightly field reports on the voice mail. "I'm at Highland and Beverly, and I'm still hearing it." "The signal drops out at Café Tropical." "I'm just around the corner but can't get you at my house."

One of the DJs tells me there's another illegal, low-power radio station operating on the same frequency as KBLT. It's called KSCR, and it broadcasts from the University of Southern California, downtown. Larry and I didn't hear it when we first went on air because it was Thanksgiving and KSCR wasn't broadcasting during the holidays. KSCR returned to the airwaves in January, when the students came back to school, and now our signals are running into each other.

It's funny that in a city that is 450 square miles, its two pirate radio stations are located only 6 miles apart. Even though we're both low power, our signals still cross and clip into the outer edges of each other's coverage areas. I don't like this situation, but I can't do anything about it. I'd move to another frequency, but there aren't any others available. The dial is too packed.

KSCR aside, the reasons why KBLT is crystal clear for miles on some nights and fuzzy on others is one of the great mysteries of micro radio. As Larry says, "All it takes is the fog rolling in or a bird sitting on the antenna, and the problems start up." Some of it is weather, but a lot of it is KBLT's antenna. Larry has helped restore it to its pre–New Year's Eve height, but unless we build a radio tower in my driveway, the antenna will never be tall enough to send the signal floating over all of Silver Lake.

It's time for a little lo-fi promotion. I mock up a flyer, using a pig sticker I found in a children's book on farmyard animals and branding its flank with the KBLT call letters. The frequency and nightly broadcast time are just below its dangling udders. I enlarge the image and make a handful of color Xeroxes, posting them in hipster hangouts—places that are frequented by music lovers but anathema to the FCC, like Millie's Diner, Amok Books, and East Side Records. I try to be the phantom poster girl when I do this, slipping into the stores, hanging the flyer in a matter of seconds, and racing away on my bike before anyone notices.

When Jasmin comes in for her Tuesday night show on KBLT, she tells me Bean, of KROQ's *Kevin and Bean Show,* was at East Side Records over the weekend and asked her about the flyer. She played dumb. Bean's parting words: "It's probably just a couple of bored kids."

A couple? Try a small but growing army, thanks to the shit he's pushing.

* * *

KROQ is the only modern rock station in the city. It is also one of its highest-ranking stations overall, part of the Infinity Broadcasting empire, one of a handful of corporations that are making bank now that the Telecommunications Act of 1996 has passed. Until last month, when the bill went into effect, a single entity could own no more than four radio stations in a single market and a total of forty nationwide. Now, they can own eight in a single market with no cap on overall national ownership. The FCC claims the Telecom Act is to promote diversity, but the reverse is happening. Huge broadcasting corporations are swallowing up littler companies, pushing marginally profitable formats like jazz and classical off the air in favor of formulaic hip-hop and modern rock.

There isn't much opportunity to change this even on L.A.'s supposedly public radio stations. Unlike the Bay Area, where the two college radio stations include both students and non-students, L.A.'s college radio stations have no such rules. I know because I tried to volunteer at two of them—KCRW, broadcasting out of Santa Monica College, and KXLU, at Loyola Marymount. I was rebuffed by both.

KCRW is a college station in name only. It may operate from the school's campus, but several KBLT DJs have told me that students are discouraged from getting involved as anything other than grunts. On-air spaces are reserved for NPR commentators and DJ professionals who've done their time at other stations. When I called and asked to get involved as a programmer, I was told I didn't stand a chance. Ditto at KXLU, but for a different reason. To be a DJ at KXLU, you have to have been a student. These closed-door policies only strengthen my sense of purpose with KBLT.

CHAPTER 6

Working at a music magazine has its perks. Countless CDs float into *Option* and *UHF* every day, and there are loads of duplicates and rejects. Scott lets his staff pick through the pile and take whatever they want, a policy I take advantage of whenever he's out of the office. A lot of the music is delivered well in advance of the release date because magazines take so long to print—a situation I exploit on KBLT, premiering records long before they are given to radio.

Beck is the cover boy for *UHF*'s July/August issue, and I get to interview him. It is March, three months before his highly anticipated follow-up record is due to hit the streets. The label sends me a cassette copy of *Odelay*. The tape is supposed to help me prep for my interview, but I make it dual purpose. The day it's messengered to my office, I air it on KBLT. Unknown to anyone but me and maybe three listeners, it's the record's radio debut.

It isn't the first time KBLT benefited from my access to

white-hot records long before they hit the streets, but is
by far, my greatest triumph. The boy is an indie rock Elvis;
his new record is so highly boogie-able I can't take it out of n.
tape deck. I fall in love with its rubbery beats and laconic double vocals on first listen.

Beck and I live in the same neighborhood, so his manager
arranges our interview outdoors on a sliver of grass between a
basketball court and a dog park just off Silver Lake Boulevard.
The interview's at 11 A.M., but I'm early. I'm nervously fidgeting
with my tape recorder to make sure it works when Beck and his
manager park their pickup and walk over. His manager introduces us, then wanders away, pretending to be interested in the
bark of a nearby tree. We have a half hour. Beck lies on his side,
harvests a blade from the lawn, and chews on its end. In his
snap-button western shirt and jeans, he's a dead ringer for
Farmer John. He even has the demeanor of an Iowan, speaking
in a slow drawl that betrays his L.A. roots.

I run through my list of questions in an effort to unlock the
complicated mind of a budding rock star for *UHF*'s nascent
readership. I warm up the conversation by heaping enough
praise on the new record to keep his ego high and flying for
days, segue into a few toss-off questions about *Mellow Gold*,
then get into it.

"Where do you seek inspiration?"

"I learn from things that are flawed. To me, bad music is
more inspiring than a masterpiece just because something
that's perfect is invisible, it's canceled itself out."

The boy is not only cute but smart.

"If you could format a tour for yourself, how would you
do it?"

"I'd do a flatbed tour of unemployment offices and Ross
Dress for Less."

I'd pay money to see that one.

"What changes have you had to make as a result of your fame?"

"None. I don't get recognized."

"Do you listen to the radio?" I have to ask. He lives around here.

"I love listening to the beat stations, Power 106 and all that."

Wow. He has absolutely no idea that, as we speak, *Odelay* is on permanent replay at a pirate radio station only two miles away.

Beck might not, but a couple of weeks later I learn that his record label does. Rudy, a KBLT DJ who also works in the industry, tells me he heard through the grapevine that Geffen (Beck's label) knows about it and will turn us in to the FCC if we don't cut it out. It's an unsubstantiated threat, but I have no way of checking it out and don't want to risk it. I remove the tape from the studio and pop it in my bedroom stereo.

A couple of months later, when I tune in to KROQ, the DJ is promoting the debut of *Odelay*'s first single, "Where It's At." If they only knew.

The DJs are whining about the station's wimpy signal, so I pass the buck and whine to Larry, who offers to come to L.A. for a weekend and raise the antenna, though there's a limit to how high we can hoist the thing without it toppling into the telephone wires.

We take a Saturday afternoon and, doubling up on Larry's Suzuki, head to Orchard Supply Hardware. Once again, we're in the plumbing department investigating pipes, only this time we're looking at three-inch-diameter steel, not the one-inch copper stuff that is bowing from the weight of the antenna and that wiggles in the wind. The steel pipes weigh about fifty pounds apiece and are normally used for high-pressure water lines, but what's good enough for contract plumbers is good enough for KBLT. We buy two twenty-foot poles, a joint to connect them, and a couple of dozen heavy-duty U clamps to attach this monstrous contraption to my neighbors' porch. It takes two clerks with weight belts to carry them to the checkout counter and then outside.

Larry plans to ride back to KBLT on his bike but waits for me while I call the cab that will carry me with the poles. I've specifically requested a van cab, but it is clear once it arrives that the cargo hold isn't big enough. The driver takes one look at the pipes, then floors it out of the parking lot. We are so screwed. I don't know anyone with a truck big enough to carry these things, and while Larry and I are, admittedly, somewhat out of our minds, we are not insane enough to try this. Larry suggests we walk them the two miles back to my house. Sure thing, big guy. I can barely hold up my end of a single pole. I suggest we hitchhike and begin scanning the parking lot for any car large enough to hold them.

Amid the shiny new pickup trucks, there's a dumpy maroon cargo van. I wait for whoever's driving it to come out of the store. It belongs to two guys. Both of them are sketchy, but I don't know what else to do. I walk up to the one who looks most likely to have showered within the last seventy-two hours and ask to bum a ride. I'm sure he thinks I'm a streetwalker. Why else would some random girl in hot pants and go-go boots come walking up to you in a parking lot on Sunset Boulevard? I mention the poles.

His bemused expression begs for an explanation. They're for an antenna, I tell him, making it clear that I'm with Larry, who will trail us on his bike. Surprisingly, he and his friend agree.

They help to load the poles, I climb in the back, and we're off. They even help carry them to the back of the house when we get home.

Larry and I spend the rest of the afternoon dismantling the old antenna mast, then straining to position the new pipes in time for the first DJ shift at six. While we're at the back of the house doing this, my upstairs neighbor, to whose porch we are attaching the poles, is on the sidewalk out front. "Susan," he yells, cupping both hands to his mouth for maximum volume. "Are you trying to reach the moon?" He doubles over in laughter at his own joke and continues down the street.

I just wave and smile. I can't believe that in the six months I've been running the station here he hasn't figured out what's going on.

I nearly bust off a mirror splitting lanes through traffic on my way from work. I'm trying to get home before six, since I forgot to leave the key under the dog dish this morning. Kerry is sitting on the steps in front of my house when I pull up and back my bike to the curb. She's smoking. The requisite six-pack of Rolling Rock is by her side.

"Sorry!" I say, ripping off my helmet and rushing for the door. "You've been waiting how long?"

"Just since morning."

We head inside, and I offer to take her sixer to the fridge while she ducks into the studio to prep for her show.

"Paige?" she calls from the front of the house, her voice skittering across the hardwood. "Did you sell a bunch of CDs?"

"I hardly have enough to play, let alone sell," I yell, pulling open the refrigerator door and depositing the beer.

"If you didn't, there's a bunch missing."

Closing the refrigerator, I notice that the window over my desk is open, and papers are scattered on the floor. I never open that window.

"Kerry? I think someone broke in here."

She runs to the back of the house. I suspect the transmitter was stolen. We head into the bedroom. The transmitter's still there, but the Hein Gericke motorcycle jacket I'd left on my bed isn't. We check the rest of the house to see what's missing.

The bathroom is closest. Someone's taken a shit in the toilet. In the studio, the headphones and the telephone handset are missing, and there are huge gaps in the wall of CDs, which are alphabetized by artist.

"I don't know if I've watched too many bad TV crime dramas, but it looks like they pulled all the CDs out in a pattern," Kerry says. "Let's see if it spells anything!"

She begins scanning the CD racks. The *B*'s are gone, she notices. So are the *O*'s and *X*'s.

"Oh my god, Paige, it really does spell something. I can't believe this. It spells 'box.'" She bursts out laughing, holding a nail-bitten hand over her mouth. "If this weren't such a tragedy it would be hilarious."

"Box, as in cunt?" I feel as if someone just swung a sack of concrete into my ribs.

"This is scary, Paige. It seems like a warning, not a random theft."

She brings up a recent *LA Weekly* article about a pirate radio operator in South Central L.A. who had just been murdered. Someone forced the guy to drink Drano. I never heard it while it was on the air, but one of the KBLT DJs tuned in once and said the DJ was encouraging listeners to "kill whitey." The station's call letters were KLLR, so maybe his death was to be expected. But I'm still scared. Maybe this is the murderer's way of asking us to get off the air. Maybe I should shut down the station. It isn't worth death by drain-cleaning product.

"I don't know about you, but I need a drink. Want one?" Kerry asks.

She pulls a couple of bottles from my crickety red fridge with the missing freezer door, hands me one, and fishes in her bag for her Camels.

I take a swig. "Do you think we should call the police?"

"I don't know, Paige. You're calling the cops into a total contraband environment." She drags on her smoke.

Her cigarette looks good. I bum one and light it with her Zippo. "We could tell the cops it's a recording studio," I suggest. "What are the chances a cop's ever seen the inside of a radio station?"

"Oh, next to none. But you need a phone to call."

The headset for mine was stolen, so Kerry goes home to get hers. While she's gone, I notice that both the broken CD player that was on the couch and the postal-worker bag I was sewing into a backpack are missing. The two working CD players, tape decks, mixing board, and microphone are all still there. The thief could have wiped us out, but didn't. I suspect whoever did it is a junkie looking for easy cash. I know all about that from living with Will.

Kerry gets back, and we make the call. While we're waiting on the cops, I call a bunch of record stores to ask if anyone has sold them about fifty CDs—all of them beginning with the letters B, O, or X. I feel like a crank caller. None of them have, so I give them my number, and ask them to let me know if anyone attempts it. They feign concern, but I'm sure they forget the conversation as soon as we hang up.

A sandy-haired man in blue rings the bell. He's alone. We walk him to the studio and show him the CD racks, then take him to the back of the house to look at the window. He's taking notes. We steer him clear of the bedroom, where the transmitter is propped behind the door, and hope he doesn't ask about the cable running up the hall from the studio and past the broken-in window to the antenna.

He's oblivious. All he does is dust for prints, warning me there probably won't be any. He points to twin holes at the bottom of the sill where the guy jammed his screwdriver and lifted the window. The thief equivalent of the zipless fuck, his fingers touched nothing but the tools. The cop leaves me with a copy of the police report and the promise that he'll call if anything comes up.

As the rumble of his V8 engine fades down the street, Kerry and I don't know what to do with ourselves except the usual when we're together: get drunk and listen to music.

I flip the transmitter switch and bring my bedroom clock radio into the studio so we can hear the show, since both speakers were also stolen. We tune it to 104.7 and turn up the volume as high as it will go. She spins Creeper Lagoon and Touch Candy as if everything were normal, but off mike we theorize about whether the theft was an inside job.

I usually leave the house key outside so the DJs can let themselves in if I'm not around, but for some reason I took it to work with me this morning. If I'd left it, I might have a better idea who broke in. If it was a DJ, he might have used the key and saved himself the hassle of prying open the window.

I think back to the previous weekend and the two guys who gave me a ride from the hardware store. They were creepy to begin with, but after they dropped me off, they just idled at the curb for a bit, watching Larry and me. It could have been them, but I'll probably never know.

* * *

Silver Lake may be the coolest neighborhood, but it is not the safest. My house is no exception. There are no bars on the windows. Any moderately skinny arm can slide through the mail slot next to the front door, unlock the dead bolt, and walk on in. I know because I've done it.

The more I think about it, the more I realize what a moron I am to keep the key hidden under the flimsy plastic dog dish the previous tenant left on my back porch. Even if it wasn't there to concierge the burglar into my humble digs, it's a disaster waiting to happen. I upgrade to a safe disguised as a can of Liquid Wrench Super Lubricant. It looks real but has a hollow center you access by unscrewing the bottom. At twelve dollars, it's a cheap improvement over the dog dish but could be just as useless. If one nefarious soul sees the DJ pluck the key from the can, the gig is up.

No, I need something foolproof. Inspired by the hexagonal signs dotting nearby lawns, I call ADT Security Systems and arrange to have an alarm installed. The servicemen place sensors on the front and back doors and a motion detector in the hallway that can "see" into both the studio and the dining room. Installation is free, but they nail you with a two-year-minimum service agreement and monthly payments of twenty-five dollars. Just two days ago, I replaced the broken CD player, bought new speakers, and picked up a phone, spending more than three hundred dollars in less than thirty minutes. Bankrupt at age twenty-nine. It's a possibility.

I call a meeting with the DJs to explain that there is now an alarm system on the house. None of them likes the idea, but they understand it. For the most part the burden of turning off the alarm will fall to the first DJ of the day, I tell them, but everyone needs to know how to use it in case the house is empty when they come in for their shows. After entering the house, they have thirty seconds to make it to the keypad in the hallway and punch in the code.

If, for some reason, they punch the wrong keys or otherwise trip the alarm, ADT will call my house. This is the one case where DJs are not only allowed to answer my phone but will have their arms broken if they don't. If they neglect to pick up, the alarm company will sic the police. If the DJs flip out and forget what to do, no worries. Detailed instructions are in an envelope under one of the turntables in the studio. Got it? There are nervous nods around the room.

The following week, ADT calls me at work. It's about noon, a couple of hours before the station is scheduled to go on air. None of the DJs should even be there. The alarm is going off. Do I want to check it out or should they dispatch one of their cars? I'm on deadline. I ask them to look into it and call back. Nothing seems amiss. The cats probably triggered it by jumping onto the windowsill, they tell me.

About a week later, I come home during *The Roky Manson Show*—the Public Enemy/Oasis/Who/Portishead/Sly and the Family Stone extravaganza spun by the hyperactive carrot-topped DJ Chris Carey. He tells me he accidentally triggered the alarm but forgot to answer the phone when ADT called. The police stopped by, but he took care of it.

It isn't too long before yet another DJ sets it off by accident. Even though I explained what she needed to do to stop the alarm, she comes in the back door, triggers it, and takes off running down the street, leaving the front door wide open in her wake. She doesn't call to tell me. In fact, she never comes to the station again.

I'm left to deal with the mess. ADT has, once again, summoned the police to my house. One guy. One gal. A cute cop couple, with props. They've got guns. Their flashlights are on. And they want to come inside. I put up a fight. I know they're trying to help but . . . I have friends in town, I say, giving them some mumbo jumbo about loaning my friends the key but forgetting to tell them about the alarm. The cops stand straight-faced, arms akimbo. Could they be more cliché? I don't think they have the right to enter, but they insist. My lack of spine does not bode well for any future FCC confrontations. Okay. All right. By all means, come on in, I tell them. Just keep it quiet because my out-of-town friends are musicians and they're recording in the studio, I lie. They walk through the house, spotlighting every room, but decide everything's fine and leave. My impending heart attack is averted.

That's it for the alarm system. We're not using it ever again. I'd rather have some stranger break in through the window, take a dump in my toilet and steal a bunch of CDs than deal with this nightmare.

The girl who triggered the alarm was new to the station—one of several I've added in the last couple months to fill two new layers of shifts, from 4 to 6 and from 6 to 8 P.M. It seemed each DJ had told at least two friends about the station who told at least two more friends, and so on and so on and so on.

We've been on the air eight months. In that time we've lost only a few DJs. One of the station's only girl DJs moved to New York. A guy who also had a show on KXLU, the Loyola Marymount station, dropped out because he was scared he might lose his FCC license if anyone found out he was also deejaying at KBLT. Their spaces aren't hard to fill. Already, there's a waiting list of people wanting to get their own shows.

In addition to me and the original two shifts of DJs, we now have Chris Herrera, the ACLU's director of public education, masquerading as Italian soccer stars Fabrizio Ravanelli, Paolo Maldini, and Guiseppe Signori for a show that centers on jazz, lounge, and rock. There's also Sulli, keyboardist for the band Possum Dixon, kicking up his heels to play vintage country. Wing, a skateboarding videomaker, is spinning jungle. Maki Tamura, another KXLU DJ, tortures everyone with a noise program. Andy Sykora, drummer for half a dozen local bands, performs unusual aural experiments under a name he lifted from a box he found in a bathroom that read 250 TOILET SEAT COVERS.

The new DJs who have already been to the station as guests are usually trained by their friends, but if they haven't been to KBLT, I do the training. Jay Babcock, a copy editor with an affinity for art rock and Brit pop, doesn't have a direct connection to the station. He is referred by a friend of one of the DJs.

It's a Saturday morning, midsummer, and I'm wheeling my bike down the driveway in shorts and slippers to get it out of my neighbor's way when I spot a lanky bleached blond knocking on my front door. Protruding from his knee-length, pea-green shorts are stick legs, anchored with beat-up Chuck Taylors.

"Are you Paige?" he asks from my stoop.

"That would be me," I yell over the engine. I park between cars on the street and head for the stairs.

"I'm Jay," he says, smiling shyly and extending a pale, skinny arm for a handshake that is surprisingly firm considering the circumference of his biceps. It's about as thick as my wrist.

I twist the knob on the front door and show Jay into the studio, the walls of which have been slowly but steadily accumulating promotional stickers, rock posters, and notes the DJs have left for one another. I introduce him to Mitzi, the little gray cat who's sleeping on the turntable. Ralph, my second cat, also wanders in to say hello.

He pets the cats and checks out my CD collection, then steps over to the DJ controls. "So this is where it all happens?"

"Yeah, except that the receiver burned out last night. The latest station casualty."

Jay picks up the pair of headphones on the table. The cord has been ripped from the left ear.

I sigh and roll my eyes. I hadn't seen that yet. "Okay. The second-to-last casualty. For some reason, the DJs keep rolling over the cord with the chair and breaking those. I have to buy a new pair every other week."

Jay laughs.

I shoot him a look. "Really, it's not funny. It's costing me a fortune."

"Sorry."

"That's all right," I say. "Just never do it again!" I point a finger at him jokingly.

"Okay. Okay!" He holds up his hands in mock surrender.

"What that means for you is that I won't be able to train you today. Too much shit's messed up. Just come to your show a little early this week, and I'll train you then."

We move to the living room and continue talking.

"You know, I've talked to you before," Jay says, taking a seat in the overstuffed chair opposite the couch where I'm sitting. Now that we're sitting face-to-face, I notice he bears an uncanny resemblance to Ed Norton, though I have no idea how old he is. With his peaches-and-cream complexion, shy smile and cocker spaniel eyes, he could be anywhere from twelve to twenty-six.

"What do you mean?"

"I called you a few months ago because I wanted to do an article about your station for this local music magazine I write for—*Strobe*."

"Oh yeah. I told you no, didn't I?"

"You said you weren't doing press until you got busted. I specifically remember the 'doing press' part. I always thought that was weird. Are you in the media? Only people in the media talk that way."

Damn. Another fissure is forming in my alter ego. "Yeah, actually. I work at a little magazine out in Santa Monica."

"*Raygun*? *Bikini*? *Option*?"

"Do I have to say?"

"No, I was just wondering."

Jay tells me he's a copy editor for two of Larry Flynt's non-porn mags—*Sci-Fi Universe* and *Rap Pages*. We talk shop about L.A.'s magazine scene for the next half hour.

"So, is it *Option*?" he asks again.

"Sheesh, you're persistent."

"I swear I won't tell anyone, if that's what you're worried about."

I'm nervous to let any of the DJs in on where I work, but for some reason I trust him, probably because he's a fellow writer. I come clean and tell him about *Option* and *UHF* but don't give him my real name. He looks like the cat who killed the canary. I'm wondering if I made the right choice.

"Hey, do you want to have lunch somewhere?" he asks, sensing my discomfort and switching the subject.

"Sure, but can we go later? I've got to go to Circuit City and buy some new gear before the DJs come in."

"On your bike?"

"Of course. I don't have a car."

Jay offers to drive. I accept, hoping to avoid a reenactment of the last time I was at Circuit City. I was there to buy a new CD player, since one of the ones in the studio had stopped working, but it fell off my bike on the way home and broke.

Jay drives the ugliest car I've ever seen. It's a booger-green Toyota hatchback that was born only a few years after he was. The interior is shit brown, a shade that was either standard issue or is a discoloration from years of back sweat.

Circuit City's only a couple of blocks away. Inside, I pick up the same kind of headphones I always get from the display just inside the door and ask one of the snake oil salesmen in the home stereo section to show me the cheapest receiver he's got. I don't need anything fancy—just something with a volume control and an LED display so the DJs can see that the station is tuned to 104.7. He tries to upsell me, but I insist on a bottom-of-the-line Kenwood for about $150. Writing out my check, I hunch over my checkbook so Jay can't read my real name. I hand it over to the sales clerk with my ID, praying he won't thank Miss Carpenter for the sale and foil my thinly veiled alter ego. He doesn't.

We leave Circuit City and have lunch at a diner down the street. Jay drops me off at my house, and I ask if he wants to get together later that night. I'm going to a party with a couple of DJs from the station and invite him to come along.

Asking guys out is not my normal MO, but I like Jay. He's cute and seems sweet and sincere. I genuinely like talking to him, and we seem to have a lot in common.

Jay shows up at my house around ten, just as a couple of DJs are winding down their show with an Al Green track. Both DJs know my radio name is Paige, but I know one of them from my KALX days, so they know my real name is Sue, and that's what they call me. Before I answer the doorbell and Jay steps inside, I remind them to call me Paige. But when I introduce everyone, they mess up and call me by my real name. Even worse, every time they flub up, they burst out laughing. Jay is giving me confused looks. Not knowing how to deal with the situation, I just pretend it isn't happening.

Jay comes in for his show the following Tuesday. I train him, he plays a bunch of Brian Eno and Oasis, and he leaves. We start talking on the phone almost every night, plotting various activities. We challenge each other to a meatball cookoff and make a tentative date to try smoking pipes. I'm his new favorite person, he tells me. Within a week, we're dating. Within the month it's a relationship.

Will and I are over. We were trying to make it work for a while after I moved to L.A., but I told him he had to give up heroin if we were going to stay together. He tried during one of his visits to L.A.—sedating himself with Vicodin and sleeping for five days straight. I thought he was through the worst of it once he woke up, but when the stomach cramps, sweats, and chills set in, it was too much for him to stand. He begged me to drive him downtown to the corner of Sixth Avenue and Bonnie Brae, where he'd heard it was easy to score. He said he just needed one hit to take the edge off, and that would be the end. I believed him and drove him downtown. We got home and he shot up. He shot up two more times before the night was through. That was the end.

CHAPTER 7

Determined to get KBLT on their radios, a couple of DJs living in Echo Park jury-rig wire-hanger antennas and hang them outside their windows for better reception. It doesn't really work. Echo Park is just a couple of miles east of Silver Lake, but the station's signal rarely travels that far. Part of the problem is our low power and low antenna, but it's also because KBLT's and KSCR's signals run into each other, snipping off the outer edges of our coverage areas.

I've never met anyone from KSCR, and I don't really want to. I've received two sneering phone calls on the KBLT voice mail from KSCR DJs, asking if we have a license, saying they suspect we're illegal and threatening to turn us in. Talk about the pot calling the kettle black. They're just as illegal as we are. If I had enough faith in my diplomacy skills to know that a phone call to KSCR wouldn't turn the situation into a second Bay of Pigs, I'd try to make peace, but I don't. I'm scared I'll piss them off, and they really will turn me in.

The climate between pirates here is much different from San Francisco, where there were enough frequencies to go around, and there existed a sort of code among thieves. But in L.A., it is not a cohesive movement. The micro radio operators are suspicious of one another and avoid direct contact. The city should be large enough to accommodate two pirates on the same frequency, but only if they're far enough apart. We are, but only barely.

Yet another low-power signal threatens to block our signal from the opposite direction, six miles west. *Billboard,* the music magazine, plans to open a club on the Sunset Strip. Outside, there will be a Jumbotron—an electronic billboard that runs advertising and announces upcoming shows. The sound for the sign will be a low-power broadcast on 104.7.

If KBLT is going to stay in Silver Lake, there's only one way to stop our signal from being boxed in, and that's to switch frequencies. I ask Larry if there's any other spot on the dial we might be able to squeeze on. He thinks 90.1 is a possibility and offers to give it a test run the following weekend when he's in town to upgrade our antenna. Again.

The station is down for two days. We take all of Saturday installing the new antenna and syncing it up with our transmitter at the new frequency. Then we take all of Sunday putting it back on 104.7. On paper, the math said 90.1 would give us better range, since the nearest station operating on that frequency was about thirty miles away. But our signal on 90.1 was even worse than it was on 104.7 ten months ago, when the Santa Anas broke the antenna in half. On 104.7 we will stay.

There's one upside to this otherwise downer weekend. For some reason, when Larry puts Humpty-Dumpty back together again, he manages to tweak 60 watts from our 40-watt system.

It's more power than we've ever managed to squeeze from our measly little rig. I'm ecstatic. Maybe, just maybe, we'll be able to reach all of Silver Lake.

I'm out of the house the night DJ Santo is on the air. He's getting ready to hand over the controls to the always enthusiastic Roky Manson, when they hear the buzz of low-flying helicopters over the already cacophonous Birthday Party record DJ Santo's spinning to close his show. Police helicopters are nothing out of the ordinary. Someone's always up to no good in this area. More than likely, it's just another moron snorting blow in his Chevy who suffered the misfortune of getting caught and is leading the cops on a high-speed chase. Or is it?

The floodlights are beaming through the front window of my living room. The helicopters are hovering directly above my house. Police cars are pulling up and parking mere steps from my door. A suspicious-looking white van is parked across the street. Is it Armageddon or the FCC? Roky is convinced it's the latter. Wanting to go out like a rock star, he scrambles for his N.W.A. record to play "Fuck tha Police." DJ Santo, meanwhile, is reading the sheet on how to shut down the station. He shuts everything off before Roky has the chance to turn N.W.A. into the last song ever aired on KBLT.

They sneak out the back door and walk down the driveway, smoking cigarettes and acting casual, trying to make it look like they came from the carriage house behind my apartment, not KBLT. They avoid eye contact with the cops, and they avoid their cars. They don't want the police to follow them and get their license plate numbers. They take a hard left out of the driveway and walk a mile to Akbar, where their friend Raquel is pouring drinks. She feeds them whiskey 'til they're drunk. By the time they stumble back to the station, the cops are gone. The station is still there.

I don't find out about this little incident until later that night, when Jay and I get back from eating dinner out and find two messages on his voice mail. The first is a panicked DJ Santo: "Paige! The FCC is at your house. Get over here fast!" The second is also DJ Santo: "False alarm, Paige. Don't worry. We took care of it."

I go home to check things out anyway. No one is there. Roky or DJ Santo must have called the next DJs to tell them what was up. I power up the transmitter, but the meter only goes to 36. We lost 24 watts in the shutdown, and that translates into a weaker signal. All that work, and we're worse off than we were before. I could weep.

DJ Santo redeems himself by allowing me to attempt live remote broadcasts from No Life, the Hollywood record store he co-owns. Nearly every weekend No Life gives up its window space for performances by whatever indie rock band happens to be in town. I figure I should capitalize on my No Life connection so the bands can appear in two places at once—live at the store and on KBLT. I consult with Larry to see if it's possible. He recommends I buy something called a Gentner, a piece of equipment that can compress the sound from the mixing board at No Life so it can travel over a phone line and into the mixing board at KBLT. Larry buys it and passes me the bill. It sets me back two hundred dollars.

Maki, the KBLT noise DJ, also sound engineers the live shows at No Life. He hooks up the Gentner and we give it a test run. It's a disaster. The local rock-revival band Beachwood

Sparks are doing their best Allman Brothers imitation, but they sound as if they're playing on the moon. We'd do better if they played into a tin can and sent the music over a piece of string. Even worse is my call waiting, which I forgot to turn off. Every time a call comes in, three beeps interrupt the music. It's only slightly less intrusive than the taco truck that occasionally parks on my street and whose CB radio, for some unknown reason, cuts into our signal, drowning out our music with their requests for more tortillas.

Mid-fall, about a year after we first went on air, the transmitter melts down, and it's going to cost at least five hundred dollars (cash only) to replace. That's five hundred dollars I don't have. At this point, my lifestyle is floating on plastic, and it's not even that great a lifestyle. I have no privacy. I can barely afford to eat. And I'm endlessly emptying ashtrays and picking up bottle caps.

On $26,000 a year, I'm barely making ends meet, and that's before shelling out for all the station's expenses. Between the mailbox service, voice mail, headphones, turntable cartridges, CD holders, occasional equipment replacement, and miscellaneous stuff like electrical tape, extension cords, and record cleaner, I'm being nickeled and dimed all the way to the poorhouse. I already buy all my clothes used. I bleach my own hair. I get my hair cut every other month at Supercuts. I rarely eat out. There aren't any more corners to cut.

I can't pay for a new transmitter unless I'm delinquent on my rent. If I'm evicted, what do I do with the station? Operate it from Griffith Park, running power from my motorcycle battery? I'm too proud to ask the DJs to chip in.

Jay isn't. He calls up a handful of diehard KBLTers and asks each of them to give me thirty-five dollars, or as much as they can afford. Some of them complain about it, but when faced with the choice of handing over some cash or giving up the station, the decision is unanimous. They want their BLT. They fork over the money.

That's the way it should be, according to Jay, who collects the cash in less than a week. Watching me dole out two hundred for a CD player one month, three hundred for an amplifier the next, he's been an advocate of dues since the first day we met. He thinks each DJ should help cover station costs and work to keep the station in order by taking out the trash, realphabetizing the CDs, compiling playlists for record labels, or performing any other of the multitudinous tasks the station requires and that are currently done by no one but Jay and myself.

I like the idea of a radio co-op, but I hate the pay-to-play concept. It goes against everything the station stands for, namely freedom. I'd hoped that the DJ waiting list would be the only barrier prospective DJs would have to contend with to get a spot on KBLT. I'm trying to provide open access to the airwaves, like the FCC is supposed to do but doesn't. How free can the station be if the DJs are paying for the privilege? But my altruism can reach only as deep as my pockets, and they are empty. I don't want to charge the DJs to be on the air, but I don't want to be snacking on Friskies and living in a cardboard box either.

A while ago, I'd bought a plastic piggy bank and stuck it in the studio, hoping people might make voluntary contributions, but it yielded only pocket change. Beginning in November, I stake a lockbox to the studio wall and begin collecting dues— seven dollars a month, roughly the cost of a decent six-pack. It's an odd amount, but five dollars seemed too low considering the rate at which the DJs break the equipment, and at ten dollars I suspected I'd be making a profit.

Bill Smith, who deejays a novelty show called *Transmissions from Endor*, is volunteer treasurer, a job that consists of counting the singles the DJs have stuffed in envelopes and dropped in the slot, keeping track of who's paid, and playing the heavy with those who haven't.

Two of the DJs say they'll pay but never do, despite repeated pleas from Bill. Eventually he refers the problem to me, and I confront each. One tells me he's doing me a favor deejaying on KBLT. He gets paid to spin dance music at Cuffs, a local gay bar. He's a decent DJ—the only one who plays house—but I have as many people waitlisted for shows as I do regular DJs. The dues are only seven dollars a month. It seems so little to ask. I tell him to pay up or leave. He leaves. The other DJ says he's tired of being harassed for money by a pirate radio station, practically spitting out the word *pirate*, as if deejaying were forced labor. He quits the station, too. My only regret in letting them go is that I didn't cut these selfish cheapskates loose even earlier for guilting me about charging them when I've given up so much to keep the station going.

These vacated spots are not hard to fill. I've been so deluged with requests for shows that I've expanded the station's hours twice in the last two months. We are now on the air from 2 P.M. to 2 A.M. daily, though I have no idea how many listeners we have. We don't get that many calls. Some people have joked that KBLT is the only station with more DJs than listeners, and I'm inclined to think that's true.

One DJ has tried to trick listeners into calling by offering free Nissan Sentras, but who or how many people are listening doesn't seem to matter to most people. Every day it seems I'm getting calls from strangers working some extremely tangential connection, hoping it will win them a show as I expand the station's hours. They are friends of friends of friends of someone at KBLT—the cousin of a DJ's girlfriend's dog-sitter, the coworker of some person's brother's urologist. But however crazy the association, they must know someone because they are calling me at home and asking for Paige.

On occasion, I get messages on the voice mail, like the one from a listener named Dwight asking for an internship. I'm intrigued by his choice of words since KBLT does not have interns, just DJs and subs. I call him back. Dwight doesn't have any radio experience, nor does he have any connection to the station other than listening to it. I have a soft spot for people like this. They remind me of my attempts to penetrate KUSF in San Francisco. Whenever possible, I give them shows, even though my bullshit detector is never set too far from high. I'm well aware that the FCC could be anywhere or anyone, so in situations like this I go with my gut. And my gut says Dwight doesn't wear a windbreaker with a fluorescent three-letter acronym or drive a van with hundreds of thousands of dollars' worth of electronic tracking gear. In fact, after talking with him for just a few minutes, it is clear he doesn't really understand what KBLT is. Despite a fuzzy signal that is likely to fade unpredictably, rampant profanity, and a tendency by some DJs to fuse Bulgarian folk music with the Rolling Stones, he does not suspect that KBLT is illegal. In fact, he thinks it's a real radio station. I correct his misimpression as best I can and, because I plan to add yet another layer of shifts from noon to 2 P.M., invite him in for a trial run the following week.

It's Tuesday morning, about eleven-thirty, when he arrives via limo. I ask what's the deal. The car service is an advertiser, he tells me—one of two companies he's lined up to help KBLT bring home some bacon. I reiterate that KBLT is unlicensed. We don't run ads because it would increase the likelihood of a bust. He seems to understand. I train him on the equipment, then leave him alone to do his show, telling him I'm at my desk in the back of the apartment in case he forgets how to turn on the microphone or cue up a record.

I'm writing in my office and listening to Dwight's show on the radio by my desk. He plays Hugh Masekela, Ike and Tina Turner, Funkadelic, Iceberg Slim. Nice show. We make it official. I assign him a shift Tuesdays noon to two.

The following week he again arrives by limo. If the limo company is driving him here in exchange for advertising, it has to stop, I tell him. He says it isn't. He just doesn't have a car. I'm not sure I believe him. I leave him alone but keep an ear on him at my desk. When it comes time to announce what he played, he is practically whispering as he gives a shoutout to his limo driver and talks up some fudge company, the name of which I can barely make out. Limos? Fudge? I have no idea how this guy came up with such freaky-ass stuff, or why he's lying to me about it, but it has to stop. I walk into the studio and ask why he is blatantly breaking the rules.

"Don't the DJs want some free fudge?" he asks.

He is incorrigible. Leaving the studio, I notice a stack of envelopes on the arm of my couch. All of them have my return address. They are addressed to the *Los Angeles Times*, *LA Weekly*, *New Times*. What are they for? I storm back into the studio, my face in a hideous twist. They're for listings in the various papers around town, he says. He does not understand that sending these announcements is like a crack dealer buying space on a billboard.

He is obtuse, he is selfish, and he is putting the entire operation at risk. I retract his show, a verdict he only grudgingly accepts. I close the door, watching him step into the limo. For the next few weeks I am terrified that he will turn me in, until I realize Dwight doesn't understand enough about radio to know whom to turn me in to.

CHAPTER

Scott pulls the plug on *UHF*, and I'm out of a job. I've always known that *UHF* wasn't turning a profit—that Scott was subsidizing its existence with ad sales from *Option*, hoping the little magazine I worked for would eventually be in the black. But that wasn't happening. Instead, *UHF*'s been bleeding *Option* dry and threatening to kill both magazines. Scott had to let it fold.

UHF isn't the only little L.A.–based magazine to go under. On November 14, 1996, *RIP*, the metal magazine published by Larry Flynt, also dies. Unlike me, who does nothing following *UHF*'s premature demise but mope around the house, *RIP*'s editor holds a wake. Jay works for Flynt and is good friends with one of *RIP*'s writers, so he's invited. He invites me, but I decide to stay home and cry.

Jay's surprised to find ex-Minuteman Mike Watt at the party, holding court in the kitchen and spieling to an audience of rapt fans. He takes the opportunity to invite the punk legend to KBLT to spin records from back in the day. He agrees.

Watt is at the station a week later. He arrives, ironically enough, in a white van. If it weren't for the bullet holes and stickers on both the driver's and passenger-side doors, it's exactly the kind of van I would suspect is the FCC and stare at from my stoop until it drove away. But one look at Watt and you know he isn't a fed. He's "flyin' the flannel" and cradling a wrinkled Tower Records bag brimming with decades-old vinyl—albums by the Who and Creedence, singles by Television, Wire, Pere Ubu, Gang of Four.

I drag the overstuffed chair from the living room as far into the studio as it will go so Watt can have a seat. It's too fat to fit through the door, so Watt has to lean forward every time he talks to Jay, who is sitting in the driver's seat and spinning singles on demand, prompting spiels from Watt after each song. Watt talks about playing the "thud staff" with the Minutemen and using his infamous "piss bottle" while on the road. It's my introduction to "Pedro speak," the colorful and poetic dialect Watt's cultivated in the forty-some years he's lived in San Pedro, about an hour south of the city.

Jay averages one or two voice mail messages per show, but tonight he gets more than a dozen.

I'm panicked about finding work now that I've been cut loose from *UHF*. Working full-time as an editor and running KBLT on the side didn't leave me any time to cultivate a freelance career. Now that the magazine's gone, I have to start from scratch. I write a couple of band profiles for a local music magazine, but they pay only forty dollars apiece. I also begin writing for the *Los Angeles Times*, thanks to a contact I made while working at *UHF*, but the assignments are rare and don't pay very well. I've got to cast my net a little wider if I'm going to make ends meet.

There's a saying: Write what you know—and that's exactly what I do. As a station operator, I'm well aware of pirate's subversive appeal. But as a writer I also know its journalistic value. The micro radio movement is classic David and Goliath. The little stations that could fight big bad radio and win—at least for now. Add in youth and music, and the story's a slam dunk. I like the idea of writing about pirate. It gives me the chance to meet other station operators and, in the name of journalism, call up the FCC and learn their latest policy toward micro radio. It also helps me understand the scope and history of what I'm involved in. I capitalize on my insider status and pitch the idea to three magazines: *Rap Pages, The Source,* and *Jane.* I haven't written for any of them, but each assigns me articles.

Jay is instrumental in getting me the *Rap Pages* assignment. He knows the editor there because it's part of the Flynt empire and because he's now writing for the magazine. I've heard that South Florida is a hotbed for underground radio, with as many unlicensed as licensed stations. Jay spreads the word to his editor, and *Rap Pages* flips the bill on a plane ticket to Miami.

There are supposed to be more than a dozen stations in the area, though I've been in touch with only one—Ghetto Radio in Fort Lauderdale, a contact I got through Dunifer when I interviewed him for my *Source* story. I trust I'll be able to find others by doing a little amateur triangulation in my rental car and by working the boys at Ghetto Radio for their connections. I suspect that, like me in Los Angeles, they either know or know of other operators around town.

It's a Friday night in February when I touch down at Miami International, sign out my rental car, and drive into the steamy darkness toward Ghetto Radio. Heading north on I-95, I seek and search my way around the dial, hearing hip-hop, reggae, calypso, Latin pop. Entering Lauderdale, I lock it on Ghetto Radio's frequency, 90.9, knowing I won't be able to tune in the station until I'm within a few miles of it. I figure I'm getting close when the dashboard speakers blast an overmodulated, "Aw shit."

Driving through the neighborhood, past the mom-and-pop chicken joints and liquor stores with bars across their windows, I begin to understand how the station got its name. I haven't seen a white person for miles, and I'm feeling eyes. I pull up to the gate of a dimly lit apartment complex and tell the tall, dark, and humorless security guard that I'm here to see some people in apartment 508. They're expecting me, I say. He closes the glass door to his booth and phones for approval. I don't hear anything but guess the guard is describing me because his lips are moving and his eyes are looking me up and down. With my blond braids and nervous smile, it would be hard to mistake me for a cop, but I guess you never know. I pass inspection. He doesn't slide his door open again, just lifts the gate and waves me toward a building. I park and go through a second security check in the foyer.

This is pretty genius, I think while waiting to be fetched. Even if the FCC figures out where Ghetto's signal is coming from, the field agents won't be able to get to it without the DJs knowing, and in that time they could flip the switch. Radio station? What radio station?

A young guy in camo pants and matching hat emerges from the elevator and lets me in.

"You Sue?" he says, throwing out a hand. I create a little handshake confusion as I fumble for the right grip. "Skitch."

Skitch is Pillsbury soft, with wire-rimmed glasses and a scruffy beard that frames his face. He leads me upstairs and down a long, beige hallway that is abnormally quiet for a Friday night. Except for apartment 508. He taps the door with his knuckles, then twists the knob. I don't see but smell the blunts as I'm welcomed into an apartment more minimalist than my own. There's nothing in the living room but a couch and a TV. I've got them beat by a bookshelf.

Skitch introduces me around. General G, the straight-faced station manager, is on the sofa watching TV. He lifts a paw in greeting. DJs Boom, Rednut, and Ladie Most Dangerous, the only other girl in the room, are with him. In the studio, Ant Man is working the turntables, headphones saddling his backward red baseball cap while he stands at the controls nodding to the beat. They all look like they're in their twenties.

The radio station is in what would normally be a bedroom, but instead of a mattress there are turntables. Substituting for a chest of drawers, a makeshift shelf with a triple layer of equipment: receiver, CD player, tape deck. The blinds are drawn. A single unshaded lamp throws some light on the white walls, which are almost entirely unadorned except for a Born Jamericans poster and a few album jackets taped near the door. Ant Man's crate of vinyl is on the floor, a 2Pac record angling out for easy access.

I hang out in the living room for a while, cross-legged on the brown shag, chatting. I'm trying to get my bearings and form at least a loose bond before turning on my tape recorder for interviews. When one of the DJs offers me a beer, I take it. I even bum a bidi from Ant Man, but I'm nervous. Pirate radio is about all we have in common. I'm three thousand miles from L.A., but culturally it feels like a million. In ten minutes I drain twelve ounces and smoke my little cardboard cigarette down to a nub.

It's time. It's getting late, and I don't want to be driving around here after midnight. I pop a cassette in my Panasonic and press Record.

In the studio, between blunt hits and swigs of beer, the DJs boast about breaking new artists and playing releases so far ahead of the commercial stations that they and the rest of the neighborhood are sick of them by the time they come out. They brag about advertising concerts on the radio that draw hundreds of people within days. They claim Ghetto hasn't had any trouble with the police because the cops are tuning in. They say they have ten thousand listeners, and I believe them. The phone's ringing off the hook.

It's clear the DJs here are celebrities, if only for a few blocks in all directions. They are full of swagger and boasts. For specifics about the station, I go to the General, who bought all the equipment and pays the rent. Our interview is alone and in the living room, where he sits on the couch while I sit on the carpet.

"Where did the transmitter come from?" I ask, wanting to suss out if Dunifer's at all connected with the Miami scene.

"It's from around," he says, his voice pure Barry White, his expression blank. His arms are crooked on the back of the sofa. Even sitting, he is tall. He scares me.

I rephrase the question, make it more vague. "Was it built on the East or West Coast?"

"The shit just appeared. I don't know where it came from. I just know how to make the shit work." He's answering me as if I were a cop, and it's making me uncomfortable.

Skitch told me that General G and everyone else with underground rigs in these parts bought them from an engineer at one of Miami's commercial stations. You just pay him a couple of grand, and he not only builds it but hooks it up. I'm asking the General because I want to cross-reference and confirm the information, but the guy's not giving anything up. I can't blame him. I wouldn't either. Ghetto's his operation.

My braids also aren't doing me any favors. In L.A., my plastic hair is hipster cool. Here, I just look like a poser moron. I'm self-conscious but press on.

"How did you choose this spot to operate?"

"Very carefully. We try to stay as secure as possible, being that what we do isn't exactly legal, so we have to protect ourself."

"Do you think this is a safe environment?"

"If anybody's comin' to see me, they better let me know. And if it's not the right people comin' to see me, we won't be here when they come to see us."

"Are you scared of getting busted?"

"No." I predicted the answer before he said it: The man's practically oozing testosterone.

"Do you think you'll get busted?"

"We're ready. We've got lawyers, money. We're ready for anything. You're not planning on snitchin', are you?" When he leans his head to one side and fixes me with a dark stare, I understand the question wasn't a joke. He's serious. I want to ask where he got the cash but think better of it.

I try to deflect the accusation by telling him about KBLT, but the way he's looking at me makes me feel like I'm lying. I'm going to finish this interview fast and sprint to my car.

"Do you want to go legit?"

"After a while. Once we got everybody locked. Once we get all the listeners that we need, then the money's there to keep it running."

"You don't think they'll be checking into your history and see that you've been operating this station?"

"They won't be lookin' into my shit. It's gonna be Joe Blow, perfect clean record, never committed a crime in his life. It's gonna be his station."

He elevates from the couch and heads for the kitchen for another beer but doesn't bring me one. From my vantage point on the carpet, his head appears almost to hit the ceiling. He takes a seat, and I hear the sigh of the can as he punches down the tab.

"We're gettin' ready to boost our wattage. Right now it's 40. We're going to 110. We got to. We got people to reach," he says, before taking a long sip.

"When are you upping it?"

"We'll call you."

I kill the tape. This hasn't been the most productive interview, but it will have to do. I cut my losses and split for the night.

It's late morning and I have no idea where I am, just that I'm driving on a wide boulevard lined with palm trees somewhere in downtown Miami. My radio is on, the volume up. I've left it tuned to 90.9, figuring there are probably other underground stations on the same frequency, just like in L.A. The dial in major metropolitan areas is so congested, they have to. Bingo. Some DJ somewhere is spinning old-school hip-hop: Grandmaster Flash, Kurtis Blow. I know it's micro radio because there's a hint of hum to the signal. Another telltale sign: The sound is mono and tinny. If it were commercial, the bass would boom.

I drive back and forth on the same stretch of cement trying to figure out if I'm getting closer or farther away from the station, but just when I think the signal's getting stronger, it blacks out. So much for my triangulation plan. I pull over and keep listening, hoping the DJ will announce the station's name and phone number. He does.

I call the World-famous 90.9 Allstars, not sure anyone will pick up, but someone answers. "T. Spoon," says a voice that matches the one I was hearing in my car. I appeal to his ego up front, hoping it will guarantee a yes later on. I compliment him on his show, introduce myself as a fellow pirate and journalist, and ask if we can meet. He seems tentative, but agrees—as long as we get together somewhere other than the station so he can see I'm for real before taking me there. He suggests a Kentucky Fried Chicken and gives me directions. Two hours later, I'm sitting in his Jeep blindfolded.

No, it's not a kidnapping. He asked that I wear one to keep Allstars' location a secret, and I agreed. I wouldn't have done it if I felt threatened, but for some reason I don't, even though T. Spoon's as big as Shaq and looks just as intimidating. His eyes are heavy-hooded, and he's wearing enough gold jewelry on his fingers to front the teeth of an entire neighborhood. Under ordinary circumstances, I wouldn't want to meet this person in broad daylight, let alone in an alley late at night. But half of being a journalist is instinct, and I get a good hit off him.

We drive a couple of blocks and park. He kills the ignition, opens my door, and leads me by my elbow toward a building. It isn't a very good blindfold. If I raise my chin and peek below the crease, I can see we're in a small parking lot behind a building. We head for a steel door, which he unlocks. Once inside, I'm unmasked. The room is Florida glamorous with a mirrored ceiling and carpeted walls. The DJ controls are immediately inside the door, between a birdcage and a tropical fish tank. The lights are low, the beats are loud, and the phone is ringing. We are the only people here.

T. Spoon picks up the cordless and settles into an office chair in front of the controls he had on autopilot while driving to get me. The call is on air. "Hey, baby, my name's Visa," the woman says, all come-hither.

T. Spoon doesn't miss a beat. "I wish I was the Discover card, baby doll, so's I could come by and see what's goin' on over there. Now whatcha wanna hear?"

She plays coy. "Whatcha wanna give me?"

He gives her "It Takes Two."

The sexual innuendo is unbelievable, but this nasty-as-you-wanna-be vibe is rampant on the underground stations here. That's part of the reason they're so popular. Subtracting out the FCC rules, they're the only stations that can play unedited rap in all its shit-talking, motherfuckin' glory—and air the shoutouts from teenagers who are so foul-mouthed in describing their sexual exploits they make Lil' Kim sound like Celine Dion.

Miami underground radio has a long and raunchy history dating back to Uncle Al, a local bass artist whose booty music has spawned thousands illegitimate children. He started Bass 91.9 in 1993. That lasted about a year before mutating into Flavor 91.9—a hip-hop station Luke Campbell of 2 Live Crew ran out of a bank building. Rumor has it the station was busted because Biggie Smalls, Genius, and other major hip-hop artists were circumventing Miami's commercial stations to be on the underground, though it might have been the commercials that brought them down. Flavor was running fifteen minutes of ads every hour, pimping everything from tattoo parlors to nightclubs. The FCC snipped its cable in 1995. By then, word about pirate was out. The legendary Sugarhill DJs jumped on 88.3, DJ Raw squeezed on at 91.7, and the Bomb set it off on 90.9. None of those stations are around anymore—most of them were raided by the FCC or imploded because of internal policy disputes—but all of them have been replaced with other underground stations. I've noticed, from the black DJs I've spoken with, that they never call it pirate. That must be a white thing.

Thanks to introductions by Skitch and T. Spoon, I visit two more operations before leaving town: Miami's Transmit Authority—a 50-watt hip-hop station run from the twenty-ninth floor of a five-hundred-unit apartment building downtown—and Peace In the Hood, a station that broadcasts from the back of a convenience store.

Miami's pirate scene is radically different from the ones I experienced in the Bay Area and in L.A. I'm shocked there are so many stations and that they have so many listeners. I'm surprised so many of them air commercials. And I'm impressed with the security of their operations. I also begin to understand another side of the micro radio free-speech issue, that pirate radio is the only on-air venue for unedited hip-hop. But as a journalist, it's hard to get a grip on this slippery scene. It's difficult to distinguish fact from fiction when everyone I talk to claims to be the first, the biggest, the most influential. In the end, I feel I've gotten the story—and the chance to see an entirely different side of pirate—but it's a relief to return to KBLT territory.

The *Jane* magazine article is just as nerve-racking but for a different reason. I'm writing about KBLT, something I hadn't wanted to do unless and until the FCC came in with a battering ram and dragged me from the transmitter, screaming and clawing for the controls. But *Jane* has me in a bind. The magazine hasn't yet launched, I've never written for it, I've already rejected one assignment—a monthly car column—and the pirate story is the one idea out of the ten I suggested that Jane Pratt likes, even though I said the subject was off-limits until KBLT was kaput.

I desperately want to write for *Jane,* not only because it's a major magazine but because it's got sass and spunk. From what I've read, it sounds an awful lot like what I wanted to do with *UHF*—make a witty, opinionated, cutting-edge magazine for women that was both stylish and street smart. When I was running *UHF,* I'd proposed the idea of turning it into what I called a "*Details* for women," instead of the fuzzily angled youth and culture magazine that it was, but Scott didn't go for it. Then *UHF* went under. At about the same time, rumors started circling that Pratt was starting a new magazine that would pick up where *Sassy* left off. Jay helped me track down the publisher, and I wrote a letter to Pratt requesting the senior editor position. Okay, so I was a little overly confident, considering how little experience I had. The job went to Bill Van Parys, a former senior editor at *Rolling Stone,* but at least he called to see if I wanted to contribute to the magazine.

My desire to write for *Jane* combined with my need for income is equal to my desire to keep KBLT running, so I strike a bargain. I'll write the story if I can use a fake name other than Paige Jarrett, lie about what city I'm in, and forget to include the station's call letters. They go for it. The article is slated for the magazine's debut in September.

CHAPTER  9

The pirate radio movement is not only going strong, it's an unstoppable force on the FM dial. Word is out on the micro radio movement, thanks to major media draining their inkwells to run stories on the subject. Four years after Dunifer first flipped the switch on his transmitter, about a thousand of us are on the air around the country—some spewing left-wing politics, others spinning music. The feds have chased a handful but are leaving most of us alone until the constitutional issues raised in Dunifer's case are resolved, and no one knows when that will be. It could be years. The FCC seems to move in for a bust only when it receives complaints or is taunted.

In the year and a half KBLT has been on the air, a few more pirates have jumped on the bandwagon in Los Angeles. Almost all of us are on the same frequency, but we are far enough apart and at low enough powers that we're not really interfering with one another. The guy who set up the Jumbotron rig for the Billboard club is running a station out of his house in Burbank, five miles north of KBLT. He doesn't use live DJs but plays CDs non-stop with a CD changer. A coffee shop in Culver City, about twenty miles west, airs poetry readings and live blues bands. Radio Clandestina, a political Spanish-language station, operates out of Highland Park, about ten miles northeast. The one station that isn't on 104.7 is called Ajax. It's also in Silver Lake, and on a much higher hill, but it shares a frequency with a nationwide religious network and has to be careful when it broadcasts so the licensed station doesn't figure it out and turn them in.

KBLT has been lucky to escape FCC attention for a year and a half, but you never know when the words of some foul-mouthed DJ might make their way to the puritanical ears of a born-again Christian who calls D.C. and prompts a bust. If that is ever the case, I plan to mount a defense similar to Dunifer's. The problem is, I don't have an attorney. I ask Chris Herrera, the DJ who works at the ACLU, to see if one of the lawyers there will take on our case. I want to be prepared if anything happens. Chris talks to one of their attorneys, but she says we won't win because we don't have a solid legal argument. The ACLU declines to take us on. I speak with a few other attorneys over the phone and meet with one in person about defending the station in the event it becomes necessary, but they immediately lose interest when I ask if they'll do it for free. This is not San Francisco, where lawyers by day slip on chaps by night and thus have a soft side for countercultural causes. This is L.A. If I do find an attorney, I'll need to pay. The thing is, I can't. I'm still without a regular job, have zero savings, and my credit cards will soon be valuable for only one thing—prying open locks.

KBLT needs a legal defense fund. Kerry volunteers to organize a benefit concert. I call a meeting to rally troops for the cause. I'm a little nervous about doing something so high profile, but almost all the DJs think the benefit is a good idea and pool their contacts. Laurel Stearns, a petite DJ who wears baby-doll dresses with men's sock suspenders, says she's friends with someone at Capitol Records. Maybe she could get Mazzy Star to play for free. Ditto for the Dust Brothers (producers for Beck and the Beastie Boys) and Polar Bear (a group with Eric Avery, former bass player for Jane's Addiction). Miwa, who works at Grand Royal, says she'll find out if Money Mark (the Beastie Boys' keyboard player) will participate. Kerry offers to call Mike Watt, who is just beginning to deejay at the station, to see if he'll play with his new band Banyan. She also offers to put a call in to Touch Candy, one of my favorite local bands, to see if they'll do the show.

Everyone, including Mazzy Star, says yes. I'm stunned. What I'd thought would undermine KBLT's indie cred is becoming one of its greatest assets—music industry support. America's discontent with the radio status quo is much deeper than I'd imagined when I first started KBLT. It taps deep into the record industry itself.

Laurel has been deejaying for six months, but I know almost nothing about her. I know that she pedals her bike all the way from Hollywood to Silver Lake in the dark and faithfully comes in to spin noise, prog, and retro every Tuesday at 10 P.M. And I know she has a lot of friends, because she never deejays alone. The girl is extremely well connected, but I guess that's to be expected when you work as a band manager.

Taz, a well-respected poster artist, is among the people she knows. He agrees to design and donate one hundred posters for the show. The posters show a crazy-eyed pig smoking a cigar, clutching a can of beer, and ranting into the microphone.

Some of the DJs who work at record labels get their companies to donate CDs and records for a raffle. Miwa gets Money Mark to sign some of his vinyl. Kerry, who is now an A&R assistant at Geffen, swings two posters and six copies of Sonic Youth's *Sister* autographed by the entire band.

Kerry had been working at Geffen Records only a couple of months when she recruited Mark Kates to deejay at KBLT. The legendary A&R agent who signed Beck and Elastica and who also worked with Nirvana, Sonic Youth, Hole, and White Zombie is such a huge fan of the station that he arranges to have a few dozen KBLT T-shirts printed on Geffen's dime. I arrange to have six navy windbreakers emblazoned with an FCC logo on the back. For the girls, I also have about fifty tank tops printed through one of Jay's UCLA friends. He runs a small silkscreening business and gives us a steep discount. The tanks come with matching panties that say DELICIOUS.

The image on the T-shirt is a slightly demented-looking little girl with pigtails and freckles holding a pig branded with the KBLT logo. It was designed by Jay's friend Camille Garcia, an extremely talented artist with a dark sensibility and a socialist disposition.

I use Camille's image to print up 150 tickets on hot-pink card stock at Kinko's. I save 50 for the door. Dave Sanford sells the other 100 through his record store for nine dollars apiece. We sell all 100 tickets in two days, with no advertising other than word of mouth and on-air announcements.

The benefit is on Friday, March 21, 1997. Four hours before the show's supposed to begin, there is already a line down the sidewalk outside Hollywood Moguls, a cave of a club with black walls and alcoves. About a dozen of us are here early to organize the T-shirts, tanks, and posters in one room. Fred Kiko is prepping his film projector to screen *Pump Up the Volume* in another. Some of the DJs, who are scheduled to spin fifteen-minute sets between acts, are already at the club and testing the equipment.

Kerry sets up the green room with alcohol. She and a couple of her friends at Geffen cashed in the $150 of Seagram's products employees get for free each year because the record label is owned by the booze giant. She stocks the room with liters of gin and vodka and cases of beer—enough alcohol to send the bands to detox after the show is over. All the bands except Mazzy Star. Their contract requires us to supply several bottles of a specific brand of Merlot. Kerry, who is used to working with bands who are lucky to score ten drink tickets in exchange for their shows, jokes about buying them Blue Nun instead. The contract also requires vegetarian meals for the band members, but the group's most difficult demand is its guest list. It's almost 50 people long. The club only holds 250. Between the club staff, the 50 DJs, and the 20 band members who are all getting in free, and the 100 tickets we've already sold, we'll be lucky to have 25 tickets left for the door. The line outside is easily over 100. So far, we've made $900 in ticket sales and paid out about $300 in expenses. If we don't up the door price, we're just having a huge out party for ourselves. I feel shitty about jacking the door price to $15, but we didn't do all this work to break even.

Within minutes of opening the front door, we're completely sold out. There's more room in the club—we aren't at capacity—but we can't let anyone else in because we're obligated to hold a certain number of spots for the Mazzy Star guest list. We can let people in only as others leave. Just then, Perry Farrell—dazed, glazed, and dressed in a white and purple pinstripe suit—walks to the front of the line with a few friends and asks to come in for free. In L.A., it doesn't matter if it's a benefit. Celebrities expect to be comped and almost always are. Their star power is too bright to turn away. KBLT is not immune. Farrell is an alt-rock blue blood. Of course we want him there, but we can't afford it. The DJ working the door asks Farrell for twenty dollars but gets more. Farrell pops fifty into the jar and walks in. His entrance is like the parting of the Red Sea.

Rumors have been circulating among the DJs that a Jane's Addiction reunion might happen here tonight, since Eric Avery and Steve Perkins are both in bands that are playing the benefit. Three of the band's four members are in the house. The DJs who are working the front lines keep their eyes peeled for Dave Navarro, but he never shows, putting this wildly inflated and highly unlikely rumor to rest.

Polar Bear plays a brief four-song set to a room that is still filling up. Their performance is followed by the one Dust Brother who showed up for the gig, then Money Mark, who performs Deep Purple's "Smoke on the Water" on kazoo. Later in his set, when he shocks his lips on the mike, he quips that it must be "pirate equipment." Watt's band Banyan is up next, to be followed by Mazzy Star.

While Banyan is thrashing away on stage, Hope Sandoval's manager is asking Kerry to escort the melancholy singer/siren to the bathroom. Kerry already has her hands full stage managing the show, but she walks Sandoval through the crowd, safely delivering her to the toilet, then returning her backstage. The entire time Sandoval says nothing to her.

A few minutes later, the band takes the stage. Sandoval wanders lazily to its center with a facial expression that's a match for her music: spaced out. Standing there in her black go-go boots and floral miniskirt, maraca in one hand, mike in the other, she stares into the distance and slowly begins her set—a cluster of songs from the band's new record, *Among My Swan*, and some oldies but goodies, "Ride It On" and "Mary of Silence," concluding with "So Tonight That I Might See."

It's a stunning show, even if my judgment is impaired. I am completely drunk. All night I've been dipping into the beer bin backstage to calm my nerves. I happen to walk into the green room moments after Mazzy Star has stopped playing. Enabled by drink, I walk up to her to introduce myself and thank her for helping KBLT. "Oh," she giggles. "I didn't know what this was. They just told me to be here." I feel like a total fool.

Earlier in the night, I'd made a similar mistake when I drunkenly assaulted Perry Farrell, thanking him for supporting the station. He gave me a look like, Listen, lady. I'm just here to see the bands. The experience gives me something other than a hangover to cringe about the following morning while I tally figures. We clear about three thousand dollars, which seems like a lot of money until I do the math on potential legal fees. Good attorneys cost anywhere from a hundred to four hundred an hour. Three thousand doesn't buy us a lot of time, but, hey, it's a start.

A number of reviews run over the next several weeks—the *LA Weekly* prints a half-page write-up of the show with pictures of Hope Sandoval, Perry Farrell, Nels Cline, and some guy in knee socks and golf shorts. The punk 'zine *Flipside* runs the review from the perspective of not getting in and disses KBLT for not having enough tickets and jacking up the door price. *Rolling Stone* also reviews the show. A couple of weeks earlier, Laurel, of the never-ending connections, had asked my permission to tell the magazine about the benefit. Nervous but thrilled that they'd even be interested, I said yes—as long as the writer didn't give any specifics about the station, either its name, frequency, or neighborhood. About a month later, a blurb about the show runs alongside a picture of Hope Sandoval, who stares blankly from the page, next to photos of Keith Richards, Wyclef Jean, and Sting. As I'd requested, none of the station's vitals are mentioned.

CHAPTER 10

Sean Kamano, a DJ who is so laid back he might as well be dead, needs a sub for his Sunday night down-tempo electronic show *Dark Lumination*. He invites his boss, Reprise Records president Howie Klein.

Klein isn't the only industry bigwig to stop by the station. Groove Radio founder Swedish Egil and KCRW DJs Liza Richardson and Trish Halloran have also been in to spin discs. True music lovers, they are attracted to the station for what it represents: absolute freedom. KBLT is the only station in L.A. where DJs can play anything and everything they want.

Like so many other substitute DJs, Howie has never been to the station, so he needs quickie instructions on how to operate our primitive equipment. It's usually the outgoing DJs who are the unwitting trainers of newbie substitute DJs, many of whom know nothing about the studio except its address, but tonight it is me.

Howie takes to the controls like the pro he is, having been a

DJ since the seventies, when he lived in San Francisco, spinning records at KUSF and KSAN. He knows music like nobody's business. In the sixties, he booked Jimi Hendrix, the Doors, the Grateful Dead, and Jefferson Airplane to play at the State University of New York, where he was a student, long before they were popular. After moving to San Francisco, he was key in ushering in the New Wave era, booking the Ramones, Blondie, the Dictators, and the Nuns at clubs like the Mabuhay Gardens and signing acts like Romeo Void and Translator to his own 415 record label.

I leave him alone to do his show—a fantastic mix of old-school underground San Francisco classics, unreleased Reprise records, and rarities. He plays Baby Buddha covering the Tammy Wynette classic "Stand By Your Man," and an alternate recording of "Safe European Home" by the Clash. Howie was with the Clash in the studio when they recorded two versions of the song. One became the first track on the group's 1978 record, *Give 'Em Enough Rope.* The other was handed to Howie. It's an original, and he's the only one who has it.

Periodically, I check in to see how he's doing. "Are these for me?" he asks, pointing to some index cards one of the DJs left all over the studio with questions and comments, like "I saw you riding the bus today" and "Do you like coffee?"

I laugh. "No. That's just one of our wack-job DJs." They're always adding little things to the studio—stickers of their favorite bands, notes for other DJs, head shots of no-name actors, vinyl that looks better on the wall than it sounds on the turntable.

Howie claims to be having a great time, even if he looks like he's dying at the controls. His eyes are red. His nose is running. He is sneezing nonstop. He's allergic to my cats. Leaving my house, he says thanks, that he'd love to DJ again. But unless I get rid of Mitzi and Ralph, that won't be happening. It is Howie's first and last show on KBLT.

It isn't just the industry that's getting involved with KBLT. It's the artists themselves. More and more bands are coming in to the station to hang out, spin records, or play live—local bands like Pop Defect and Possum Dixon, and national indie-rock darlings like Railroad Jerk, Girls Against Boys, Speedball Baby, Barbara Manning, Ben Lee, Donna Dresch, and the Dandy Warhols.

At this point, KBLT is on the air fourteen hours a day with a regular staff of about fifty DJs, including myself, and a list of at least thirty others waiting to get in. It's taken more than a year and a half, but the programming is finally beginning to break free of its indie rock mold as more people with specific musical tastes and vast libraries join the station. DJs like Bennett, who painstakingly pieced together the unreleased Beach Boys' record *Smile*. Or Paul Greenstein, who brings his 78 player to the station so he can spin vintage swing, jazz, and bop.

Still, there are some indie rock holdouts. Flipping through the studio logbooks to see what the other DJs are up to, Jay is appalled that so many of them are playing bands that already get a lot of airtime on college radio and commercial modern-rock stations like KROQ. In Jay's opinion, KBLT's mission is to air music that isn't played anywhere else on the FM dial. He thinks that radio should be tied to its community—that it should uplift its listeners by airing music they might not know they even want to hear. I agree, but not completely. KBLT's mission is also to let the DJs play whatever they want, and if that includes some of the bands that are popular on other stations, so be it.

He posts an anonymous notice on the bulletin board asking the DJs to stop playing the Beastie Boys, Beck, the Chemical Brothers, DJ Shadow, and the Jon Spencer Blues Explosion. His note lasts precisely one week—long enough for the DJs to lob accusations of "dick-tator" and scribble all over it with various expletives. By the time Jay returns for his show the following week, the note has been crumpled into a ball and pinned to the wall.

About 90 percent of my records—the ones I owned before starting KPBJ—have been stolen. CDs by George Jones, the Beastie Boys, Public Enemy, PJ Harvey, the Velvet Underground, Charlie Parker, Wire. They're long gone. So is vinyl by the likes of Big Black and the Shaggs. I wish the DJs would leave as much music behind as they have taken, but I've inherited very little by default, and those records I have gotten are absolute crap—"Ms. Pac-Man," for example. It's hardly a fair exchange.

CD theft is a slap in the face to both me and the handful of other DJs who actively acquire music for the station. It's a lot of hard work to keep up the flow and stay current. Each month, DJ Santo pores over the DJ logbooks to compile a list of records that are getting the most play. He then mails and/or e-mails that list to the labels that are sending us music. Every third day or so I ride over to our mailbox to pick up the booty—between three and ten CDs each day. The records are then logged, labeled, and added to the new-releases bin, only so people know what's come in, not because they have to play it. The CDs stay in the new bin for about two weeks before they are filed into the main CD wall, which is arranged alphabetically but not by category, having been replaced by whatever else has floated in.

A lot of what we get the DJs won't touch with a ten-foot pole, but the most popular stuff is inevitably stolen, usually within days. I can understand how a DJ might accidentally throw one of the station's CDs into his bag, but as soon as he figures that out, you'd think he'd bring it back next shift. Some do. Most don't. Others, I suspect, nick the CDs as payback for collecting dues or "borrow" them until the record is officially released and they can buy their own copy.

I put KBLT stickers on the jewel cases, thinking it will help stem the tide of CDs drifting out my door. DJ Santo and I write "KBLT" all over the cover art and the actual CDs as a reminder that, no, these CDs do not belong to them, regardless of how many times they play them. None of this works. I leave friendly notes on the bulletin board, asking the DJs to please check their bags before they leave. Still, the CDs walk. My stickers and notes become increasingly stern, even threatening: "Steal this and I'll kill you!"

They know it's an empty threat.

I hate playing Cagney and Lacey. A benevolent dictatorship is much more my style, but that doesn't work when people don't play nice. Because it's the new CDs that are stolen most often, I begin numbering them, making the DJs sign a form vouching that all the CDs are there when they leave. Finally, there's some improvement.

The tiny nuclear family living directly upstairs from me, and whose porch is a crucial part of the station's infrastructure, moved out a while ago. For two glorious months I didn't have the stress of worrying about falling into their laundry line when we did repairs, or the ever-present fear that the studio noise was bleeding up through my ceiling and into their living room. In the eleven months the station had been on the air, I was surprised my neighbors never figured out the math: that an antenna + dozens of people carting record crates = illegal radio station. But they didn't. At least, they never let on. If they did know, they never seemed to be bothered by it. Then again, they did decide to leave.

For a brief period, the family was replaced by Christopher Scot, a magician with a pet pig named Porkchop. The chances of someone owning a pig for a pet were already low, but the chances of that same someone moving in next door to a radio station named after a bacon sandwich were pretty phenomenal. Life is sometimes so surreal.

Until Christopher and Porkchop settled in, the most interesting people in the building had been Sara—a clothing designer with sharp teeth and hair black as ink (she's the neighbor with whom I share my entire interior wall)—and Dennis, the Vietnam vet with a missing index finger whose floor doubles as Sara's ceiling. Christopher and his porcine friend upped the building's quirk factor by a mile. If Christopher wasn't wandering around with a ventriloquist's dummy, he was practicing the accordion. And he was just as likely to be wandering around in a top hat and tails as he was to be wearing jeans and a wife beater tank top. Meanwhile, Porkchop made his home in the parking lot directly behind my bedroom window, so that in addition to falling asleep with the radio station in full swing, I was waking up to an extraordinarily hungry animal squealing for his breakfast. At 5 A.M.

Despite the farmyard alarm clock, Christopher and I hit it off immediately. Even so, I didn't tell him about the radio station. Two weeks later, I learned he was a regular listener. Once he realized the bass coming from my apartment matched the beat on his radio, he put two and two together and knocked on my door. He even comes by when I'm not around. One time he delivered ice cream sandwiches to some DJs who had mentioned on air that they wanted them.

It was an enormous relief to have a devout KBLT listener living directly upstairs from me, but the apartment wasn't ideal for Christopher because the only place to put his pig was on the concrete out back, and that wasn't good for Porkchop's little hooves. So when the people living in the carriage house in the back moved out, Christopher switched apartments so Porkchop could roll around in the dirt yard.

Good for Christopher and his pig. Bad for me. I've been extremely lucky with my neighbors so far, and that luck is just about to run out.

Coming home from Jay's one morning, I find stacks of boxes littering the sidewalk in front of my house and my new neighbors—a couple, about my age. I pray they are cool or, even better, deaf. Steve is a short soft body with wiry hair and glasses. His girlfriend Catherine is a transplanted Londoner in a Donegal cap and bell-bottoms. She's got an accent and an apartment in the hippest neighborhood in L.A. so she thinks she's the shit. She cops attitude when I introduce myself, and Steve follows suit. I'm 0 for 2.

That already pitiful ratio plunges into the negative about a week later when I realize my new neighbors have ripped out their carpet to get at the hardwood floors. It might be an aesthetic improvement, but it's also a noise nightmare—for both of us. We might as well be living in the same space. Their footsteps echo throughout my apartment, and, though I can't make out exact words, I can hear their voices through our shared ceiling/floor. Making the problem even worse is that our apartments are identical in layout. Steve is a part-time community college teacher who spends most of his time at home. He sets up his office in the space directly above the studio.

At this point, the station is on the air from noon to 2 A.M. every day. Despite my repeated pleas for the DJs to keep the volume down, most of them ignore the note on the receiver advising them that the maximum allowable volume is 40, a warning that I've repeated in a bold-lettered sign posted on the wall just two feet in front of the DJ chair. I tell the DJs that if they want to play the music loud, they'll need to wear the headphones, but even I know that isn't fun when you've got company. When I'm not around, I suspect my rule goes unheeded. Their attitude: This is pirate radio. There shouldn't be any rules. I truly wish that were the case, but the DJs are not the ones who have to suffer the consequences directly. When they pack up their records and go home, it's over.

For them, but not for me. Steve is obviously annoyed by the noise but only "talks" about it with his feet, skipping rope in his living room and jumping up and down on the floor above the studio, sometimes screaming at the top of his lungs. I understand his frustration. I'd be upset, too, if there were a crazy radio station running full steam ahead in the apartment below. But as annoyed as he is with me, I'm just as aggravated by him. I could complain about a lot of stuff—how the two of them drive in and out of the driveway in the middle of the night and wake me when they walk up the wooden stairs behind my bedroom window, how sometimes when I'm in the bathroom taking a shower I overhear, courtesy of the window well, slaps and sighs that conjure up images I'd rather not think about. But I'm not about to bring that stuff up. In fact, I'm terrified to attempt a conversation with Steve about how to resolve our differences. He seems so hostile. And I can't move out. The station is too entrenched.

The best I can do is continue harassing the DJs about keeping the volume at a reasonable level and install more soundproofing. I drop two hundred dollars on foam tiles and staple gun them to the studio ceiling. If Steve would install carpet, our problems might be solved. I didn't know how good I had it with my old neighbors—a simple nuclear family who went to sleep at ten every night and never complained.

Jay and I already spend most of our nights together, sometimes at my house, others at his. But, tired of our neighbors' noise and of DJs tramping through our bedroom to turn off the transmitter at 2 A.M., we begin to sleep at his place every night.

Cake, a DJ with more than a passing resemblance to Meat Loaf, is on the phone in the kitchen, checking voice mail messages between tracks by Nirvana and Throbbing Gristle. He forwards through calls from record labels checking on the CDs they've sent, listening for any messages specifically for him. There's one: a request for Styx.

He returns the receiver to its cradle, but just as he's turning to walk back to the studio he sees some William Burroughs lookalike standing on the back porch. Cake doesn't recognize him but goes to the door to see what the guy wants, unhooking the latch and opening the door a crack.

"Can I help you?"

"Yeah, hi," the man says, looking past the 250-pound mass that is Cake and into the apartment. "I like what you're playing."

"What do you mean, 'like what you're playing'?" Cake asks. "I'm just watching my friend's place."

The man smiles. "I tracked the signal."

"The signal? I don't know what you're talking about. Like I said, I'm just house-sitting for a friend and I'd like to get back to my nap." Cake starts to close the door.

The man chuckles. "Would you mind if I came in?"

Cake blocks the door. "I don't even know who you are. Sorry."

Realizing he isn't getting anywhere, the man backs off. "Well it's a nice operation you've got anyway. Keep up the good work."

Yet another radio geek who knows enough about low-power radio to use the terrain to track our location. I wonder how long he's been looking for us, not that we're so hard to find.

So many people in the local music scene know about the station's "secret" location at this point that local bands leave their CDs on my doorstep. Anyone paying any attention to my house could piece together what's happening here. Every two hours, someone new shows up with a DJ bag and, sometimes, crates full of records. Music is throbbing fourteen hours each day. An antenna juts above my roof.

It's late morning. I'm sitting at my desk attempting to write when half of my living room ceiling crashes to the floor. I'm sure it's because of my new upstairs neighbor's incessant jumping up and down. The combination of him skipping rope in his living room and pogoing above the KBLT studio have finally proved too much for my 1940s plaster.

It is also wearing me down. I'm tired of trading dirty looks whenever I see Steve and living with the constant fear that he will not only figure out the source of his problems but how to fix it. All it would take is a single phone call to our friendly FCC field office and KBLT is toast.

The stress of living this way is sending my blood pressure through the roof. My repeated pleas that the DJs keep the volume down haven't helped, nor has the new soundproofing. I think I need to move. I post a notice on the studio bulletin board asking the DJs to be on the lookout for a new place—preferably a free-standing structure in the upper Silver Lake hills. Where I think I'll get the money to move to a new, more radio friendly altitude, I don't know. I can barely afford to live on the lump that I do.

Over the next couple of weeks, I get a couple of dozen calls from the DJs. There's a one-bedroom apartment on Descanso for $595, a small house on a hill off Riverside Drive for $650, a warehouse just south of Santa Monica Boulevard, a one-bedroom apartment on Rowena. I drive by all of them, looking to see if they're on big enough hills and in buildings where the neighbors won't get suspicious. None of them are high enough, cheap enough, or stealth enough. Relocating the station to any of them would be a lateral move at best. Not only that, switching locations would mess up our coverage area, and I don't want to sacrifice any established listeners. I decide to stay put, figuring the devil I know is better than the one I don't.

I'm back from a week-long trip to Portugal with my dad and sister, during which I picked up three bottles of the contraband *aperitivo anisado* called "absinto," made, according to its label, from "alcool, agua, acucar, destillado e corante"—alcohol, water, sugar, and something I suspect is wormwood though I'm not sure because I don't read Portuguese. Anxious to try it, the first night I'm home I crack open a bottle, offering some to Roky Manson, who is about to kick off his show and kick out the jams with the Stooges' "Search and Destroy." We both throw back a shot, then another, even though it tastes like Nyquil. We're hoping to rush the hallucinogenic effects, but they don't hit me in the fifteen minutes I'm waiting for Jay to pick me up to see a movie at the Vista Theater, around the corner from his house.

Hearing a honk in the driveway, I take off for the night, leaving the open bottle of absinthe on the floor outside the studio. In the two-minute drive from my house to the theater, all Jay and I hear is part of the Clash song "Gates of the West" before we park and head into the theater. An hour later, all anyone else is hearing is the *shlip shlip shlip* of the turntable's stylus repeatedly running into the center label of *Led Zeppelin IV*. Somewhere between tracks by Sepultura and Howlin' Wolf, Roky has worked his way through three-quarters of the bottle.

In her apartment two blocks away, Miwa is getting together records for her show and listening in. She mistakes the "shlipping" for a minimalist noise track until she gets to the station and knocks on the door. No one answers. She begins to yell Roky's name. Still, no answer. It is past ten. Her show has already started, and she can't get in.

Miwa is at the station with her brother Jim and her impossibly tall friend Fritz, who longboarded from his house to meet them. They begin to panic. Fritz goes to the side of the house, craning his neck to see into the studio from outside the living room window. "I don't see him," he calls out to Miwa and Jim, who is worming his slender arm through the mail slot to unlock the dead bolt on the front door. The three of them stream into the house, checking the studio. Roky isn't there, but the station is on and the turntable is spinning. Jim and Fritz begin searching the house, calling Roky's name, while Miwa jumps on the mike to tell anyone who might still be listening that she'll be right back. She pulls an Alec Empire seven-inch from her bag, cues it, and pushes up the fader.

"Miwa! Come in here!" Jim yells from the bedroom.

Miwa springs out of the studio and down the hall. "Oh my god. Oh my god!" she screams. Roky is passed out facedown on my bed, the near-empty bottle of absinthe by his side. "Roky!"

She slaps his face. "Oh my god, Jim. What if he's dead?" She checks his pulse. He's alive but in a deep, alcohol-induced slumber.

"Roky. Roky!" Jim and Fritz are saying over and over.

The record in the studio runs out. Once again, the needle is hitting the label.

"I'll deejay," Jim offers. He runs back down the hall to the studio, switching to Sly and the Family Stone without a mike break.

Miwa goes to the kitchen to get a glass of water, and sprinkles some on Roky's face. He is dead asleep.

"He's not going to wake up, Miwa," Fritz says.

"Yeah, it doesn't look like it."

"Roky," Fritz laughs.

"I'm going to kill him when he does, you know?"

They decide to let him sleep it off and go to the studio to finish their show, periodically checking on Roky to make sure he's still breathing. He's still sleeping when they spin their last record and the next DJ comes in for his show. Roky wakes up the next morning in my bed with the word *loser* written on his forehead in Magic Marker.

Steve Hochman, a writer for the *Los Angeles Times,* calls to interview me for an article he's writing about KBLT. I immediately want to kill him and whoever gave him my number. Sure, it's flattering that the *L.A. Times* is interested in KBLT, but a story in such a high-profile paper is sure to get us busted.

I tell him I won't participate and ask him to refrain from writing it. He says the story's already been assigned. I ask if he can leave out the station's call letters and frequency and what neighborhood we're broadcasting from. I've got a station to protect. Steve says he'll check with his editor and get back to me. His editor says no. All I can do is wait for the bomb to hit.

It does two weeks later. Responding to the thud of newspaper on concrete, I pad outside to pick up my plastic-wrapped *Los Angeles Times*. It is a beautiful, sunny Thursday morning. The air is clean, the sky Windex blue, but I almost spit out my coffee when I see, on page 14 of the Calendar section, under a banner that reads RADIO, a story titled "Land of the Pirates" accompanied by a photo of local band, Extra Fancy. Its caption not only names KBLT and its frequency, but says we are based in Silver Lake.

I read it faster than you can say Evelyn Wood, then pick up the phone to cuss out Steve Hochman, using more than a few FCC-banned expletives. It is seven-thirty in the morning. He's barely awake—most definitely taken aback by my accusations that his article is, in effect, turning me in to the authorities. Mostly he just listens, but eventually he calls me on my hypocrisy. I am, after all, advertising the station on flyers all over town, and those flyers give not only the name of the station but its call letters and phone number. He's got a point, but I'm still enraged.

It isn't long before I'm fielding calls about the story from the DJs. A handful complain about the two local scenesters who squealed to the *Times*. Jay calls to see that I'm all right, then storms down to a local store to confront one of the people quoted in the article who claimed KBLT was doing advertising tie-ins, even though KBLT is staunchly anti-ad. Kerry asks if I'm having a heart attack. I'm not, though I've got a serious case of the shivers. I am terrified to turn on the transmitter when Amos, who hosts a Latin jazz program, comes in for his show at noon. What if the FCC trails him up the stairs? What will I really do when the FCC knocks? I walk outside and look down the street, scanning it for white vans. Nothing.

I turn back inside and walk to the bedroom. I flip the switch. Let the men in blue windbreakers come to take me away.

CHAPTER 11

Kerry is sick of her show and sick of Los Angeles. She's disappointed and disgusted that KBLT has morphed from the close community of friends it started with into a loose network of strangers, many of whom she thinks are involved for the wrong reasons—not because they love music, but because deejaying at KBLT has become the hip thing to do. She quits the station and takes a job with Sub Pop records in Seattle. She'll be moving in two months.

This is devastating news. Kerry is my closest girlfriend in L.A.—the only DJ, other than Jay, I feel close to and hang out with. Kind, funny, smart, interesting, spunky, sassy, generous—she's a rare breed in Los Angeles. She's one of the few people I've met who doesn't take herself too seriously and who isn't all about her career. One of the few industry people I've met who has interests other than music, namely politics, pop culture, and boys. Who wouldn't miss a girl who sets up dates with several different guys on the same night and at the same bar?

Her move will be a huge loss to me personally, but it's also a jolt to KBLT. Not only has she introduced a lot of good people to the station, but the benefit was successful largely because of her near-magic coordinative powers and magnetic personality. Easygoing but efficient, sincere yet goofy, Kerry is that rare sort of character who can motivate people to do extraordinary things without making them feel put out, the kind of person who can play the heavy with a light touch. I'm going to miss her.

Over the past couple of months, Kerry has been subbing out her show more and more often. Usually, it's whatever boy she's been locking lips with, but sometimes it's Mike Watt, with whom she's become relatively good friends since the benefit. Kerry suggests Watt be her permanent replacement when she leaves and calls to offer him her Friday-night slot. I'm shocked he says yes, but KBLT is true to his punk rock ideals.

Watt brings his "Pedro slice" to "the big, lettuce and tomato" in June with *The Watt From Pedro Show,* "broadcasting somewhere north of the equator." Each show starts with John Coltrane before veering off into punk, more jazz, and spiels about the songs he's played. It's a living, breathing, oral history of L.A. punk—a true gem of a show, hampered only by Watt's nervousness and bungling of the controls.

Kerry trained Watt on the equipment long ago, but his show is still plagued with technical problems. Jay is usually on hand to help him with the equipment, but Watt still leaves the microphone on while the music is playing. He also forgets to adjust the speed on the turntables as he bounces from seven-inch singles to full-length classic rock records, a problem compounded by the dexterity and speed required to spin such a show. Punk singles are, on average, little more than a minute long, never longer than two. The studio is already hot, but when Watt finishes his shows, he is almost always drenched in sweat.

He is also a little toasted, drinking whiskey and chugging beers like Popeye throughout his two hours on air. The day Watt started deejaying at KBLT, as a big thank you, I dropped seventy dollars of my own money to buy him a case of Boddingtons ale, which I wrapped with a ribbon and left in the refrigerator with a note: FOR WATT. At the end of the show, I tried to get Watt to take what remained of the beer with him, but he asked me to keep it in the refrigerator for his next show.

Big mistake. The next day, the beer was completely gone, my note lying on an empty rack in the refrigerator with the question: "Watt who?" Apparently, completely forfeiting my personal living space isn't enough for some DJs. They have to help themselves to whatever I have in my house. This lack of respect for my home is becoming more and more common as the DJs' connections to the station extend further and further from the circle Quazar initially hooked me up with and as I spend more and more time at Jay's, leaving the DJs on their own without supervision.

The line between private and public has been blurring ever since. I discover globs of gum in the hallway carpet and pipe screens on my kitchen floor. That's in addition to the countless empties that I find each morning and the cigarette butts that are lost in the cushions of my couch and sometimes stubbed out on the floor. There are so few rules at the station, you'd think the DJs could at least pick up after themselves, but only some do.

KBLT has played host to virtually every local band in the area, but, thanks to a bunch of well-connected DJs, we're beginning to attract higher-profile national and international acts. Mark Kates brings in John Squire of the Stone Roses, who is in town to promote his new band, the Seahorses. Judah Bauer of the Jon

Spencer Blues Explosion stops in with his friends from Speedball Baby, another funk-and-punk blues act from New York City. DJ Santo, who also runs a record promotion business, arranges for Jason Pierce, the strung-out genius singer for neo-gospel psychedelic band Spiritualized, to stop by KBLT before his show.

Pierce is running late. His manager calls and asks if someone can pick him up from the El Rey Theater, where he's playing later tonight, and bring him to the station. Any number of DJs would gladly suffer through the hour of hundred-degree rush hour traffic for the opportunity to bask in Pierce's narcotic greatness, but DJ Santo and Roky Manson get the job. Sound check still isn't finished when they pull up to the club. At this point, Pierce is running so late the station will be lucky to have even two minutes on air with him, but DJ Santo and Roky don't care. While waiting, they are treated to their own personal performance of "Walking With Jesus" and "I Think I'm in Love" in an otherwise empty theater.

We have ten minutes from the time Jason walks through my door, waxy and bug-eyed, until he's back in the car. He isn't here so much for an interview as to spin some of the records he's bought while on tour. He gets to play only a couple from his stack, since he squandered most of his time quizzing me about the station setup and counseling me to move the transmitter somewhere other than my house, like the land-based pirates in England. He tells me the stations in London operate from abandoned buildings with cemented stairwells so the cops can't get in. I'm flattered by his concern, but I think he's being overly paranoid.

It's late August 1997. We're averaging about a band a week, but this week has been chock-full. On Tuesday, indie electronic songstress Lida Husik came in, followed by Ventilator and Six-teen Deluxe on Wednesday, Jason Pierce on Thursday, and now the Flaming Lips. You'd think these people have better things to do with their time than stop by a radio station that can only be heard for two miles in any direction, but I guess they like what we represent—freedom—as much as I like what they bring to KBLT—legitimacy.

The Flaming Lips are in town to promote *Zaireeka*, a high-concept box set that requires all of its four CDs to be played at the same time on four separate CD players. A couple of nights earlier, Jay and I had witnessed the public debut of this off-kilter symphony during a listening party at Warner Brothers Records. KBLT already has a copy, even though it isn't officially out 'til October, but it isn't getting much airtime. We only have two CD players.

Wayne Coyne, the band's eccentric and adenoidal lead man, stops by the studio with Steve, the band's drummer. It's a sunny Friday afternoon, smack-dab in the middle of summer, but Wayne is wearing a yellow rain slicker. The two are here to spin records for an hour or so before they head back home to Okla-homa. They look through the library and pick out a couple of CDs, then take turns playing Bread and Black Sabbath while Jay and I work the controls, but Wayne loses interest pretty quickly. Just like his songs and liner notes, he is curious, kooky, and chatty, chatty, chatty. He is far more interested in quizzing Jay and me about the purpose of micro radio and how the station works than he is in spinning records.

Mike Watt was just the tip of the iceberg, as far as celebrity DJs go. Don Bolles, drummer for the Germs, also has a regular show. And Keith Morris (former lead singer of Black Flag and the Circle Jerks), Bob Forrest (singer for Thelonious Monster), and Sylvain Sylvain of the New York Dolls have all subbed at the station.

I know very little about these bands. I know they're prestigious, but I've actually heard very little of their music. The year L.A. punk broke (the first time), I was living in suburban Illinois, a permed junior high schooler with an orange hair pick in my back pocket that said HOT STUFF, even though nothing could have been further from the truth. I was just beginning to get into music at the time. My favorite bands: Loverboy and the Knack. Like most other prepubescent girls, my entrée to rock was through the cheesiest of vehicles. I had no idea there was an alternate musical universe, and even if I did I don't think I would have liked it. I didn't hear of the Minutemen or the Germs for another half decade, by which time I'd transformed into an imitation-towhead mod who listened to nothing but New Order, the Cult, and Blancmange.

With all the music bouncing around my apartment these past two years, KBLT should have been my chance to catch up, to learn enough of L.A.'s musical history to at least play *Name That Tune*, but I haven't. I stopped paying attention to individual songs long ago. The music in my house is so omnipresent that it has become background. I crave silence more than sound.

I rarely listen anymore to the new music that comes into the station. I appreciate the record labels' support, but I'm getting jaded by how much awful music is committed to CD, so I'm tuning out. I used to be an indie queen, playing new alternative music almost always, but I've recently retreated into the past and reformatted my show. It is now exclusively French pop, most of it girl singers from the late sixties.

I'm not the only DJ whose show has evolved out of modern

rock into something far more esoteric and interesting. Mark Kates, who used to play a lot of Brit pop and rock, has done a complete 360. He is now DJ Carbo, a hardcore junglist playing nothing but underground drum 'n' bass. From his mad imitation of London 'ardkore pirate radio, you'd never guess the guy is married, has a kid, and lives in a mansion.

CHAPTER 12

My article about pirate radio appears in the debut issue of *Jane* magazine, prompting a flurry of congratulations among the DJs. The flurry only builds when *LA Weekly* calls KBLT the city's "best closet radio station" in its "Best Of" issue, saying it's a "musical oasis." I'm pleased that the writer compliments not only the concept but the actual programming, which has blossomed in its almost two years on air. What started as a college radio clone, albeit an illegal one, has really come into its own as more and more DJs with enormous record collections and very specific musical expertise join the station. We now have entire shows devoted to country, jazz, Latin, soul, lounge, bop, folk, punk, rock, jungle, dance, pop, kitsch, French, and gospel, as well as shows that mix any or all of the above. I couldn't have programmed the station better if I tried.

KBLT is becoming way more popular than I'd ever thought possible. Every week or so I scroll through the station's voice mail to see if there are any messages for me that the DJs forgot to tell me about. There's a call from a guy named Richard saying he loves the station. A message from Steve in Hollywood saying keep up the good work. A guy who claims to have deejayed on the legendary offshore English pirate, Radio Caroline, asking for a show. Someone named Knee with questions about the schedule. Another person asking for advice on how to start his own station. Someone complaining about the signal. And my favorite call: "If you put those records in the sun, you can make ashtrays out of 'em," he says, before asking to be a DJ here.

The station is still an underground phenomenon, but the people who know about it are fanatical. I regularly receive pig-related gifts from visitors—pig stuffed animals, pig-shaped Christmas lights, pig calendars. The KBLT promotional stickers Camille designed are showing up on tons of cars, and people are now telling me in person that they moved to Silver Lake just so they could tune in.

Silver Lake is becoming the new Seattle, or, at least, it's getting its fifteen minutes of fame. Everyone from the *Los Angeles Times* to *The New York Times* to *Details* to *Vanity Fair* are writing about it. In the past, 90210 might have been the zip code of preference, but it is now 90027. I'm sure it isn't the taco stands and tranny bars that the reporters are responding to. It's the stars. Madonna, Leonardo DiCaprio, Brad Pitt, Beck, the Beastie Boys—all of them have headed for the hills of Silver Lake in recent years.

An Associated Press reporter calls to interview me for a story she's writing, "L.A.'s Hip Displacement." I guess the fact that Silver Lake has its very own radio station makes it even hipper. I agree to the interview because it isn't about KBLT, just the neighborhood. She says she won't mention the station's call letters or its frequency. She has no idea Paige Jarrett isn't my real name.

* * *

Spaceland's owner, Mitchell Frank, calls to offer us two free pairs of tickets for every show, to be given away over the air. His club is one of the hottest in town and gets some of the better-known national acts who are popular on college (and now pirate) radio. In the coming weeks, Bis, the Beat Junkies/Company Flow, Red Krayola, and Kelley Deal 6000 are scheduled to play. I'm psyched. It's a win-win-win situation: Mitchell gets some promotion for his shows (which we already announce over the air), the listeners get to see some of their favorite bands for free, and the DJs get affirmation that people are actually listening because they're getting calls for tickets. It isn't long before other clubs follow suit—the House of Blues, the El Rey Theater, and another local indie rock club that recently started in a rundown Mexican bar a mile from KBLT, the Silver Lake Lounge.

There are some downsides to the station becoming so popular. A handful of listeners have called to say that KBLT has sold out and that they are now listening to Ajax, a Silver Lake pirate radio station upstart. It is less popular than KBLT and therefore even more pirate, they say. And while the station is doing so well that there's actually backlash, I'm not doing so great. I'm run ragged by the simultaneous pressures of making a living and running a homegrown radio station. In addition to pitching and writing stories, I'm constantly picking up music and adding it to the library, updating the on-air schedules and DJ lists, making and delivering flyers. I'm stressed, and I'm taking it out on Jay, rather than on the people who are the source of my problems—the DJs who steal the station's CDs, borrow my books without asking, use my phone to make long-distance calls, and trash my house with fast-food wrappers, cigarette butts, beer cans, and drug paraphernalia.

I feel taken advantage of. Angry, I sometimes pick fights with Jay, and we temporarily break up, a situation I immediately regret for many reasons. He's my best friend and my only true support system in L.A. One of the more unbearable after-effects of our short-lived breakups is that I'm forced to sleep at home with the station on the air almost around the clock. During one 2 A.M. program, while I'm attempting to sleep at my house, I hear a DJ singing the Brady Bunch's "Sunshine Day" at the top of his lungs. I have closed every possible door between the studio and my bedroom, and it still sounds like he's singing directly into my ear. I throw on my robe and storm down the hall to ask whoever's singing to keep it down. He says he will, but the moment I'm back in bed he's at it again. I put a pillow over my mouth and scream at the top of my lungs.

I've tried to run the station with a light touch to allow the DJs as much creative freedom as possible. While that strategy has made the time they spend at the station great, it's made mine horrible. I understand that part of the appeal of pirate radio is what it stands for. It is not only a reaction against bad radio but a reaction against authority. Whenever I attempt to impose rules, I am shat upon. Even simple rules—common courtesies like cleaning up after themselves—go unheeded.

The only possible upside to the trash are the hundreds of bottles and cans that are left in the DJs' wake each month. I'm seriously considering cashing them in for a small profit. If I had a car, I'd wheel them to the recycling center just a few blocks from my house and add the twenty dollars or so to my almost nonexistent monthly income. It isn't much, but it's something, and I am that desperate for money. Even with the DJs paying dues, KBLT's technical difficulties tend to spring up when the coffers are running on fumes.

My career has finally shifted out of neutral and into first gear—I am now writing regularly for the *Los Angeles Times, Jane,* and *The Source*—but I'm not writing for them often enough, they never pay me when I expect them to, and the rate I'm earning per word is barely enough to cover my basic expenses. Still, I don't want to get a regular job. Between the radio station and my writing, I hardly have the time, but it is becoming clear that I have to. I'd rather clean bathrooms at a roadside gas station than be a secretary again, so I rack my brain thinking of things I like to do that might also make me some money. I decide I'll be a motorcycle instructor. I go through sixty-five hours of training and begin teaching safety classes almost every weekend.

CHAPTER 13

The pirate radio movement's main bad guy gets a new face when President Clinton appoints and the Senate confirms William Kennard as FCC chairman. Part of Kennard's job will be to come down hard against illegal broadcasters, but I wonder if he'll recognize why the situation is so out of hand. More important, I wonder if he'll do anything about it. Prior to taking over as FCC head honcho, Kennard was the commission's general counsel, helping to implement the Telecommunications Act of 1996. Perhaps in the past year and a half he's seen what the act has actually accomplished. Instead of opening the airwaves to competition, all it's done is ensure that the little fish are swallowed up by the sharks. In the last twenty months, four thousand of the nation's eleven thousand radio stations have changed hands—many of them bought by monolithic broadcasting corporations like Clear Channel Communications and Infinity Broadcasting.

Kennard is an L.A. native and the first African-American to chair the FCC. In the mid-1980s, he was part of the commission's Advisory Committee on Minority Ownership in Broadcasting—a type of ownership that's becoming increasingly rare. At the same time, Kennard was assistant general counsel for the National Association of Broadcasters, the powerful Washington, D.C., lobby group that exerts an unholy amount of influence over the FCC. It's hard to tell if there's hope for the cause with Kennard at the helm.

Claudia Wilken, the judge who is presiding over Dunifer's case in the Bay Area, rejects the FCC's request that she stop Dunifer from broadcasting. Instead, the judge orders the FCC to come up with a decent argument against Dunifer's First Amendment case. Wilken's ruling buys the movement even more time, possibly as much as two years if the case proceeds to trial. It is early November 1997.

One week later, the FCC takes its battle to Florida. The FCC may be losing its case against Dunifer, but that doesn't mean micro radio is legal. At 6 A.M. on a Wednesday, a twenty-man multi-jurisdictional task force composed of FCC agents, federal marshals, a SWAT team, customs agents, and local police break down longtime illegal broadcaster Doug Brewer's front door, handcuffing him and his wife and forcing them to the floor at gunpoint. Their crime: operating Tampa's Party Pirate station without a license. For twelve hours, the task force ransacks their home, stripping it of everything related to the operation— broadcast equipment, computers, CDs, records. They even bring in a crane to dismantle the 150-foot radio tower Brewer had built in his driveway.

Brewer, aka Craven Moorehead, is a burl of a man—a tubby forty-three-year-old with long hair and a beard. He rides a Harley and runs an electronics shop. He started broadcasting in the mid-nineties, decorating the radio tower with lights and playing Christmas music for the neighborhood. By the time his station was busted, more than twenty-five DJs were involved, playing mostly rock music, much of it local.

Brewer may have been busted because he was taunting the FCC, though other pirates claim that WHTP, The Point—a commercial alternative rock station—turned him in. The Point's general manager had complained to the media that Brewer's station, on 102.1, was causing interference with his station on 102.5. But micro radio operators have another theory: Brewer's station was showing up in the local Arbitron ratings.

The Party Pirate wasn't the only unlicensed station raided on this "Black Wednesday." Two other Tampa operations also went down at the FCC's hands. Kelly Benjamin, better known as Kelly Kombat to his listeners on 87.9 FM, and Arthur Kobres, who ran antigovernment propaganda on Radio Lutz 96.7 FM, were also busted.

Good thing we're having another benefit, because chances are we're going to need the money. Our second benefit will not only beef up our legal defense fund but celebrate KBLT's more than two years on air. It's scheduled for December 9.

Our next show is at the El Rey Theater, a gorgeous deco-style club in L.A.'s mid-Wilshire district. Lida Husik (who's now a regular KBLT DJ), the Michael Whitmore Ensemble, Possum Dixon, and Mike Watt, who just released his punk rock opera *Contemplating the Engine Room,* are scheduled to perform. Local drag queen Dr. Vaginal Cream Davis will host the show. My neighbor will do a magic act. Someone's girlfriend will give tarot readings. One of the DJs will work the crowd in a Santa suit, handing out free CDs, stickers, and other swag. Kari French—a local performance artist and KBLT DJ—will make "pussy prints," rolling her cooch with paint, squatting on pieces of paper, and selling them for a buck apiece.

About four hundred people show up for the benefit. My neighbor the magician is scheduled to do a fifteen-minute performance that involves a decapitation, a shooting, a midget, a pig, and a few minutes of him dangling above the crowd in a straitjacket filled with candy. It's the biggest crowd he's ever performed for, and he milks it for all it's worth. He hogs the stage, leaving only when the crowd gets restless and begins heckling him, throwing things, and calling for Mike Watt, who is headlining but doesn't get to play until almost 1 A.M. The club closes at 2.

A couple weeks after the benefit, Watt brings his first "spiel guest," Sonic Youth guitarist Thurston Moore, to KBLT to share the hot box and rap about punk's salad days.

Jay and I are walking out of the house to get a burrito from El Pollo Loco down the street, when a sedan pulls up to the curb and rolls down the window. "Is this KBLT?" the guy in the passenger seat asks, pointing to the door we just came out of.

"Sure is," I say, confident he isn't with the FCC.

The driver kills the engine, and four guys pour from its doors. One of them looks like Shaggy from *Scooby-Doo*. He's skyscraper tall with strawberry-blond hair and marble-blue eyes.

"Is that Thurston Moore?" a guy across the street asks loudly.

It's surreal to have so many punk legends as houseguests when I barely even know their music, but it is also understandable. Like the punk movement, pirate radio is about individuality, independence, and freedom. It's about rejecting what you're given, and creating something new on your own.

It's a week before Christmas, and Eddie Muñoz, who spins an incredible sixties garage rock show, takes a break from Canned Heat and the Troggs to spin some off-color holiday classics: Arlo Guthrie's anti-FBI ditty "The Pause of Mr. Claus," the Youngsters' "Christmas in Jail," Santo and Johnny's "Twistin' Bells."

It's a micro radio version of Orson Welles's *War of the Worlds* when, during an announcement of Santa's kidnapping by aliens, the transmitter overheats and clicks off. A listener calls to leave a message. Just as the song Eddie was playing asked her to get on the space satellite and find Santa, her radio fuzzed out, then clicked over to KCAQ, or Q 104.7—the urban-format station broadcasting on the same frequency in Oxnard.

Laughter does not normally rank in my top ten of first-response emotions when disaster strikes the station. Those positions have long been held by (1) annoyance (that KBLT needs fixing—again), (2) anxiety (that the problem might be major), (3) dread (the repairs will cost a fortune), (4) sadness (the station is off the air), and (5) fear (of losing listeners). But when the transistor burns up and blows out during Eddie's show, I crack up.

I've fixed this fuzzing-out problem before, so it's only a half hour before we're back in action and Eddie picks up where he left off, continuing with a triptych of "Blue Christmas" covers by Willie Nelson, Elvis, and the Ventures. It's practically an occasion to celebrate when I can identify a problem—and fix it. But that doesn't mean I like doing it. I hate going into the guts of the transmitter, but in the last month the transistor has blown half a dozen times. On each occasion, I've had to extract the one that's been burned to a crisp and replace it with a shiny new one. The station hasn't been working so well since Larry tweaked the power. It's acting like he gave it a kidney when it was on the table for a liver transplant. But Larry can't come to L.A. for another week, so I'll keep playing triage nurse until the doctor is free.

Without Larry, KBLT would never have existed and KPBJ would have lasted only a week. From a technical point of view, Larry has been invaluable. He has devoted endless hours and more than a little of his money to make sure the stations stayed afloat, driven only by his belief in me and our shared vision of what radio could be if regular people were just given the opportunity to be on air. I am extremely grateful for his generosity. I don't know many people who, at the end of a long workweek and on a moment's notice, would ride a motorcycle eight grueling hours in the dead of night to work even more on the radio station—this time for free.

Even so, in the two and a half years we've been working together, we've had our spats. Larry is frustrated by my slave-driver demands for "Santa Monica"—a stronger, more solid signal that can reach beyond Silver Lake. And I'm frustrated because I suspect some of his "repairs" are bogus. I believe they do nothing but satisfy Larry's obsessive-compulsive desire for electronic experimentation and cost me money I don't have.

Layered on top of this ongoing feud are time constraints (station maintenance must happen as quickly as possible so the DJs can continue on schedule), exhaustion (all of us have already worked our asses off all week), and the clash of contrasting personalities attempting to achieve a common goal. Sometimes it's just too much.

A few weeks after the transmitter starts acting up, Larry comes to L.A. to fix it. It's a huge relief, but I never know when the next technical blowout will happen.

I'm relieved to get a call from Chris W. offering his engineering expertise. Of the hundreds of phone calls I've gotten from people asking to get involved with the station, Chris W. is only the second person, after Larry, who's wanted to work on the tech side. Everyone wants to be a DJ, but having an engineer who lives just a couple of miles from the station instead of four hundred means KBLT won't have to be down as long if there are problems. It does not mean Larry is out of the picture, only that I'm no longer at the mercy of his schedule and geography.

Chris tells me he doesn't know much about how pirate radio works, but he has a friend who runs a micro radio station in Tucson, and he likes to fiddle with electronics. Chris's dad owned a Radio Shack franchise and would bring home broken radios for him to play with when he was a kid.

After speaking on the phone a few times, I invite Chris over to check out the station. There's nothing for him to fix. It's mostly just a meet and greet so we can suss each other out. Chris is exceedingly thin, with wavy dark brown hair and pale blue eyes that could bore holes through wood. He is soft-spoken, mild-mannered, and polite. I have a good feeling about him.

I have no idea what Chris thinks of me. The first time I meet him, I'm a redhead. The second time, I've got hot-pink dreadlocks. *Jane* magazine has accepted another story idea of mine. I offered to cut and dye my hair four different styles in four weeks to test a theory: People's reactions are based less on a person's face than on image, part of which is hairstyle. When I meet Chris, I'm in the final stages of my social experiment—most of which I've performed on the DJs.

In a matter of weeks, I transition from a Lady Godiva blonde to a Louise Brooks bob to a Peter Pan clipper cut and then cotton-candy dreads. I've been a blonde the entire time the DJs have known me, so things don't get interesting until I embrace the dark side and go black. Taking advantage of the kiss-me-quick red lips and raccoon eyes that are painted on me for the magazine's photo shoot, I head to the Garage nightclub, a divey bar with bands that is one of my regular haunts. Dana, DJ of the *Hi-Falootin Randomonium* show on KBLT, is working the door. She doesn't recognize me even though I make eye contact as I hand over five dollars. Neither do the other two DJs I see. I bum a cigarette off one of them and the other hits on me, completely oblivious to who I am. And these are people I've seen almost every week for well over a year. It is like I've taken an invisibility pill.

The DJs already think I'm a little bonkers, so when I make the move to red, they don't even bat an eye. It's only when the dreads are woven in that the gig is up. Hipsters are notorious chameleons, but even they recognize this is a bit much. When I tell one of the DJs I'm researching a story, she says, "I hope they're paying you a lot of money."

While I'm in my pink dreadlock phase, a friend suggests I call another woman she knows who also has pink dreads so we can swap stories. Her name, unbelievable as it seems, is Laura Bong, and she uses the opportunity of our conversation to get her own show on KBLT. When we meet, I see her dreadlocks put mine to shame. Hers swing past her butt and are decorated with everything from buttons and Barbie Doll shoes to Cracker Jack charms, bells, tassels, even a dog's squeeze toy.

Her slot: eight to ten Saturday mornings. Her format: talk. Every week she comes in with two things—an extraordinarily well-sculpted poodle named Lulu and a stack of manila folders dripping with news clippings from various papers and magazines. Each folder is dedicated to different segments on her show—Social Porn (lurid news of the rich and famous), Debs Behind Bars (socialites' bad behavior), Rock Cock (which celebs are screwing whom), People Who Are Dead (recent obituaries), and the less sensationally titled Animal News and Science News. *The Laura Lee Show,* as she calls it, is the only talk program on KBLT.

A couple of weeks after Laura starts deejaying, she comes to KBLT with a copy of Madonna's "Ray of Light" single, a heavily guarded song that won't be in record stores for at least another month or so. Laura, who works at Disney, said that a woman who was hoping to be hired as the music supervisor for Disney's remake of *The Parent Trap* came to the studio lot with her hottest property—the new Madonna single. The song was on cassette—one of only a few copies Warner Brothers had released from the studio. Laura, who is working on the movie, was asked to walk the tape to the dubbing studio, watch them make the dupe, and walk it back to the director. She walked it over, watched them make the dupe—her dupe—then make another for the director. The woman didn't get hired as the music director, and the single didn't make it into the movie, but Madonna's "Ray of Light" debuts on KBLT days after it is delivered to Disney.

There have been a lot of interesting newcomers lately, filling in the station's early morning slots. DJs like Generic Eric, a Nordstrom's shoe salesman whose show is partially devoted to the girl at the Mac makeup counter, with whom he's obsessed, and Maryann Bowers, who plays all classical music for a show called *Mrs. Kunkel's Music Class.* Then there's Eddie Ruscha, who's apt to use anything and everything as musical complement. During one of his shows, I was drilling a hole in my kitchen cabinet. When I clicked it off and turned around, there was a microphone lying on the floor to pick up the sound.

By all rights, the Jesus and Mary Chain should not be visiting KBLT. They are far too legendary a band, and we are far too small a station. But, once again, Laurel has worked her bottomless pit of connections to bring us some marquee value, aided by her friend CeCe, who does publicity for Sub Pop, the band's new label. In an act of unbridled nepotism, CeCe bumped another radio station from the group's interview schedule to stop by KBLT instead.

I doubt the band will even remember being here. They are that drunk when they show up. Clearly, the brothers Reid have been drinking for hours, if not days. They can barely walk, let alone talk. Nevertheless I offer those of pasty face and limited walking ability a glass of wine after they stumble through the door. Always the hostess. Jim shakes his head no and slumps on the living room couch along with the Chain's third link, guitarist Ben Lurie, but William says yes and trails me to the kitchen. He plucks the glass from the counter only seconds after I've filled it, but before he even takes a sip he sloshes a wave of red wine down the front of his slate-gray sweater. I give him a paper towel to clean up, but he's distracted.

"What are *those*?" he asks, sputtering, slurring his accent, and defying Western standards of personal space by teetering within an inch of my nose. He reaches out to tug on a couple of my braids, which I had woven in after finishing my hair story. I had them redone earlier in the day, and they are so tight they appear to be sprouting directly from my scalp. "You're bald!"

I'm tempted to play tit for tat and run my hand through his poofy 'do, but I just laugh and return him to the studio, where

Laurel is pulling CDs from the library. William spots the empty chair and takes a seat at the controls. He picks up my *Meet the Beatles* record and, after a protracted struggle to fit it over the turntable's spindle, drops the needle.

"Ay! There's no sound!" he says to no one in particular.

The turntable isn't even spinning. Laurel leans over his shoulder, presses Play, and pushes up the fader. He lets "I Want to Hold Your Hand" play halfway, then starts to scratch the record—not moving the platter back and forth like a DJ, but literally scratching the stylus back and forth from the outside edge to the inside of the vinyl.

"Fuck the Beatles!" he slurs into the mike, sending the meters into the red.

It's the first in a half-hour-long string of off-color commentaries before the band's manager picks them up to take them to the airport. William, who has been speaking from somewhere deep in his subconscious, has to be torn from the controls. Even then, we can't pry the wineglass from his hands.

"I quite like this glass," he tells me on his way out the door. "All right if I take it?"

"Go ahead. They're Ikea," I tell him.

"Then I'm going to take it!"

"Consider it a present from KBLT."

"I'm going to take it! I'm taking your glass. I am! I am!" he says, looking over his shoulder until CeCe redirects his attention to the steps so he doesn't fall.

CHAPTER 14

I've been living in L.A. more than two years, but I still subscribe to San Francisco's free weekly newspapers, mostly for nostalgia but also to keep tabs on what's going on. There seem to be so many more interesting and creative things going on in San Francisco than Los Angeles. Reading the *Bay Guardian* one weekend in February, I'm drawn to a photo on the contents page. An extremely pretty girl is sitting high in a tree. The caption says her name is Butterfly and that the tree's been her home for two months. I'm intrigued.

I want to read more, but the paper's a tease. There's no story. Just that one picture. I call the *Guardian* to get more information. They refer me to Earth First—the environmental group for whom she is protesting the logging of old-growth redwood trees. Yes, they confirm, she really is living up there. She hasn't touched the ground in sixty-some days.

I have to write about this. I send an e-mail to *Marie Claire* magazine in England, for whom I've just started writing, and leave phone messages for my editors at the *Los Angeles Times* and *Jane*. *Marie Claire* declines, but both the *Times* and *Jane* want the story. I write articles for both, reporting from vastly different angles. For the *Times*, I interview Butterfly from the base of the tree. For *Jane*, I climb into the tree and live with her for a weekend.

Sleeping in a tree two hundred feet off the ground on a platform secured with rope and covered with tarp put sleeping at my radio station into perspective. It is February, and the extreme cold and rain coupled with the fear of falling and the deafening noise of the whipping wind and flapping tarps makes KBLT seem calm in comparison.

I rented a car to drive from Los Angeles to the tree, which is near the hippie haven of Eureka, about three hours north of San Francisco. During the nine-hour drive both ways, I listen to one thing: a cassette one of the DJs loaned me. He knows I've just switched my show's format from anything goes to sixties French pop, and the tape is nothing but. I adore this cassette. It's some of the best music I've ever heard. The DJ who loaned it to me was Don Bolles, who, being the master of mind-fuck that he is, intentionally removed the card that identifies the songs and their singers, telling me it's a secret and that I'll never figure out who they are.

Obviously, Don doesn't know me very well. Tell me I can't do something, and, rest assured, I will. Within a month of my return, I've not only figured out who they are but tracked down copies of their records. I've figured out that the best music isn't just from the 1960s but specifically from 1967 and 1968, and I dig deeper into those years, finding even more obscure gems— like Antoine et les Problemes, who cover Jimi Hendrix. My program is improving, and it shows in the phone calls. I'm actually getting a few.

* * *

Susan Bean, a producer for the CBS national news, calls. She read my story about pirate radio in *Jane* magazine and wants to bring a camera crew to KBLT for a segment she's producing about micro radio. As long as she plays by my rules, I say. She can't reveal my name, the station's call letters or where we are in L.A., nor can she film the outside of my house, the antenna or the transmitter. Bean agrees to all my demands, so I tell her we'll participate.

The morning the CBS news crew is scheduled to arrive, I'm helping the *2-Martini Breakfast* show live up to its name. Half of L.A. is still asleep, but Morgan (the program's host), Max (his guest bartender), and I are well on our way through a liter of vodka, having polished off a first round of chocolate martinis, shaken not stirred. Morgan and his guests regularly forgo coffee in favor of alcohol while spinning Rat Pack classics, but I don't usually drink in the morning. Today it is liquid courage.

CBS News is easily the biggest media KBLT has been involved in. Coverage in the *Los Angeles Times*, *The Washington Post*, *Jane*, and *Rolling Stone* hasn't prompted the FCC to knock at my door with a battering ram, but this is national television. Even though the producer has agreed to all my requests, I'm still nervous. I'm scared there's something I forgot to mention—some detail I've overlooked that might prompt an FCC SWAT team to descend on my roof, shimmy down the window well, and hold a gun to my head once the CBS crew clicks off its camera and goes home.

Jay is very much against my participating. He thinks I'm flaunting KBLT's success and asking for trouble. As long as I'm cautious, I think nothing can happen, but there's a certain amount of cockiness that comes with being on the air so long. I wonder if I'm just being a showoff. I justified KBLT's participation to the DJs as promotion of the cause, but there's another reason. It appeals to my vanity. I'm craving recognition for the sacrifices I've made.

Damn, Max makes a good martini. I'm sitting alone on the couch in the living room, giddy from downing a second drink on an empty stomach, when the doorbell rings. It's clear from the camera and microphone men that these aren't the feds, but I'm always cautious when answering the door. Susan Bean introduces herself through the glass before I unlock the dead bolt and let them all in.

Max emerges from the studio with his shaker and immediately offers them drinks. They decline. Susan smoothes her take-me-seriously-but-not-too-seriously navy pants suit and takes a seat on the couch. Her helpers ready their equipment and scatter to the back rooms, surveying my apartment for possible shots. When one of them calls "Susan" from the back of the house, we answer in unison. It happens again just a few minutes later. How embarassing. Susan just laughs and says, "You're a lousy thief." Really, I'm just a lousy actress.

Her crew will be here all day, interviewing the DJs on camera as they come in for their shifts. They film about two seconds of Morgan, then me, since my show's next. I have a hard enough time simultaneously figuring out what to play and operating the controls when I'm alone. Under the lens, lights glaring, I sweat my way through tracks by France Gall and Françoise Hardy, never once looking at the camera. I'm too self-conscious.

Susan interviews me in my bedroom after I play my last pouty-mouthed French girl. I'm sitting in my office chair, my back to the studio so the camera can capture the DJs in the background. They're getting a pretty good cross-section today, between Morgan doing his best Dean Martin impression, the tattooed and branded Divinity Fudge, and Roky Manson in his *lucha libre* mask.

Unfortunately, I'm going to miss half the fun. At two, I need to leave for *The Leeza Show*. I've never been on TV before, and suddenly I'm being filmed twice on the same day—both appearances prompted by articles I wrote for *Jane*.

"My Schizo Life as a Blond, Brunette, Red, Dread-head," the story about switching my hair style and color four times in a month, was a hit. A columnist at the *L.A. Times* wrote a blurb about it, which was read by a DJ for a local radio station who invited me on the air for an interview, which was heard by a producer for *The Leeza Show*, who asked me to appear on a segment about "hair horror stories." Things are getting weird.

Today's the day I meet Leeza. One week earlier, the show's producers asked me to put the pink dreads in again so they could pretape a segment of people reacting to my hair as I walked down the street. I'm wearing a pair of Buddy Holly glasses with a camera hidden in the nose bridge to capture their responses. The pink hair is one thing, but the glasses take the look to new heights of absurdity. I look ridiculous and feel so completely uncomfortable that I can't come up with the pithy comments the producer has asked me to sass to the camera that's trailing me. People always assume I'll be entertaining because of my articles, but it's harder to pull off in person. It's a disaster.

I'm alone on stage when they run the segment of me walking down the street. Then Leeza asks me questions that the producer has practiced with me at least a dozen times. The producer wants to make sure I won't mess up and blurt out something stupid when I've got my alone time with Leeza in front of the camera. I still feel uncomfortable, but I guess I do okay. At least I don't hyperventilate and pass out.

* * *

I feared the *Leeza* and CBS News segments would run on the same day and that someone with a lot of time on his hands and an ax to grind would see both shows, figure out Sue Carpenter and Paige Jarrett were the same person, and turn me in. The news piece airs in April, about a month after it was filmed. My *Leeza* segment won't run until August.

I watch the news segment at Jay's house. The piece is running as part of series called "Lie, Cheat, Steal." If the producer had told me that up front, I wouldn't have participated. I nearly have a heart attack when a clip of me is followed by an interview with FCC Chief William Kennard. When I either wasn't around or wasn't looking, they filmed everything I'd asked them not to: the antenna, the transmitter, the front of my house. They even took footage of the reporter driving down the strip of Sunset Boulevard right around the corner from my house. Any idiot could figure out the neighborhood from the street signs and storefronts. I could shoot the television.

I thought CBS was going to present an accurate and fair portrayal of the issues. Instead, the segment made KBLT look like nothing more than an excuse to listen to music and get drunk. Working in the media, I should know as much. When you agree to an interview, you think you can control the message with your answers, but you can't.

A few days after the CBS segment was filmed, the FCC serves a summons to KIND Radio, the Texas micro station that was also profiled in the news segment, as well as four other micro radio operators. The summons requires them to appear in Washington, D.C., for an "order to show cause," one that would explain why the FCC should refrain from issuing injunctions against them. The FCC had fined each of the operators eleven thousand dollars.

A couple of months later, Judge Claudia Wilken sinks pirate's flagship station, Stephen Dunifer's Free Radio Berkeley. Dunifer's case was built on the argument that the FCC's rules banning low-power radio stations were an infringement on free speech. Judge Wilken takes the wind from Dunifer's sails, saying he cannot claim the FCC's regulations are unconstitutional because he never bothered to apply for a license. Specifically, she prohibits "Mr. Dunifer, and all persons in active concert or participation with him ... (a) from making radio transmissions within the United States until they first obtain a license from the FCC; (b) from doing any act, whether direct or indirect, to cause unlicensed radio transmission or to enable such radio transmissions to occur." Free Radio Berkeley goes off the air immediately after the court's decision is announced.

At the time of the ruling, a spokesman for the FCC's Compliance and Information Bureau boasts it has shut down more than 200 illegal stations in the last two years. He also says the agency has a list of 112 pirate stations currently in operation around the United States.

KBLT has gotten away with so much crazy stuff, it just doesn't seem real that we'll actually get busted. We've tempted fate too many times. I can't help but wonder if KBLT will be the next station to go under, but there have been no indications the FCC's onto us. In fact, the station's doing better than ever. KBLT stickers are beginning to show up in some pretty strange places, like TV. One of the DJs told me she'd seen one stuck to a school locker on *Buffy the Vampire Slayer*. Another DJ saw one on a clipboard on the sitcom *News Radio*. At this point, we are on the air twenty-two hours a day. So many people are coming into the station that I'm going through at least one roll of toilet paper a day. The only time we're off the air is from 4 to 6 A.M., and even those shifts are beginning to fill up. Listeners continue to call, and bands continue to come into the station.

Watt is away on round three of his tour promoting *Contemplating the Engine Room*. While promoting his record, he often talks about KBLT and the micro radio movement. I've heard him discuss it on both KCRW and KROQ. Witty, outspoken, and intelligent, he's a perfect mouthpiece for the cause.

While Watt's away, former Circle Jerks singer Keith Morris fills in. It's a cool summer night when, unknown to anyone else at the station, Keith brings the Red Hot Chili Peppers— Anthony Kiedis, Flea, Chad Smith, and John Frusciante, who just rejoined the band. Their performance on KBLT marks Frusciante's debut with the re-formed lineup.

Keith's association with the Peppers dates back almost two decades, when Flea played with the Circle Jerks, but it was when Keith was hanging out at a mutual friend's house that Anthony Kiedis stopped by, and he asked if the Chili Peppers might want to play an acoustic set at KBLT. They were all over it. It's probably a relief for them to play at a place where fans aren't begging to sniff their underwear.

Keith doesn't give them KBLT's address. He never gives it to anyone. Instead, he meets them at the Circus of Books, a porn shop and magazine stand down the street from my house, and the five of them walk up the slope to the station, carrying guitars, bongos and congas, with more than a few watchful eyes following them. When they step into the station, the drum 'n' bass show *Rumble in the Jungle* is in full swing. Wing, in his cargo pants, and Paris, with his dreads, are scratching on the turn-tables and taking turns with the reverb as they shout out over the music. They throw on their last twelve-inch when they see Keith.

While *Rumble*'s packing up, the Peppers set up their gear in my hallway. Keith takes over the controls, kicking off a set with Jimi Hendrix, continuing on with Frank Black, the Soft Boys, T-Model Ford, Hüsker Dü, and Soundgarden, and anchoring it with Motorhead's "The Chase Is Better Than the Catch." The Peppers are ready.

John Frusciante steps up to the mike, which is resting on a chair at the studio's entrance, and gets things going with "I've Been Insane," playing guitar and singing while Chad Smith plays bongos behind him. Flea's up next. He sings "I Can't Remember" before the rest of the band joins in to play "Skinny Sweaty Man," "I Could Have Lied," and "Not Great Men." With only one microphone, it sounds less like a performance and more like an eavesdropping session during private rehearsal, but this is how bands always perform at KBLT. They take a break to play some of the music the band brought in—Rammstein, Betty Davis, PJ Harvey—then head into another set.

I am not around for any of this. Jay stops by to see if Keith needs help with the equipment, but I'm on a plane from Florida, where I was chasing down clues for a story about an identity-stealing murderess for *Marie Claire* magazine. I know about the Peppers' performance only because Jay tells me about it when he picks me up at the airport.

It's hard to figure out if it even really happened. Few people had heard the show. It takes me several weeks to track down a tape of it, and that was from a writer for the SonicNet Web site who had recorded it to write a review.

* * *

Jay is always inviting guests to his show. Sometimes it's a coworker wanting to play Hollywood show tunes. Other times, celebrities. During one of his semiweekly visits to Meltdown, a heavy-on-the-toys comic-book store in Hollywood, Jay meets former Misfits/Samhain singer and *Satanika* comic artist Glenn Danzig and invites him to the station. The horror-punk legend is into it even though he's never heard of KBLT.

Danzig has been living in Los Feliz for years, and his house is something of a landmark. At least, everyone seems to know what and where it is. It's the haunted house on Franklin Avenue—a rundown craftsman with a weedy unkempt lawn ringed by a wrought iron fence. I've ridden by it countless times but have never seen anyone there.

When Danzig comes to KBLT, he is dressed to match his soul, or at least the soul of his music—entirely in black. He is short, with arms like clubs bulging from his muscle tee. His smile is toothy and goofy. He is a flesh-and-blood cartoon.

Jay asked him to bring whatever he wants to play, and he has. A lot of it is trip-hop—Tricky and Garbage—and heavier industrial stuff like Godflesh. As with so many other musicians who've come to KBLT, what he listens to is radically different from what his own bands sound like.

The New Times, a free weekly newspaper, names KBLT L.A.'s best radio station in its "Best of '98" issue. I'm flattered but would have liked them to get my gender right. The writer refers to the station operator as "some random dude from San Francisco." The presumption that I'm a guy is a huge hot button for me, since, at times in the station's life, strangers have come to

KBLT and treated Jay as if he were the one running it. While Jay has certainly helped enormously with various aspects of the operation, it's maddening to make as many sacrifices as I have only to have credit assigned by gender.

I've just left Body Builders Gym after forty-five minutes of running to nowhere on the treadmill in preparation for my latest *Jane* assignment—posing for *Playboy*. I'm walking down the sidewalk when a car screeches to a stop at the curb and the driver runs up to me. He looks familiar. I think he's visited the station, but I can't remember his name.

"Paige!" he says, his eyes in a hurry. "What's that song the DJ just played?"

I'm confused. "What song?"

"That song on KBLT that just ended. Do you know what it's called?"

I have no idea what the song is, but I love the absurdity of this situation. Walking down the street, I'm nowhere in the vicinity of a radio. Does he think the station is hardwired into my brain? I suggest he call the studio and give him the number.

The gym I go to doesn't play KBLT—they like the band Erasure—but a lot of local businesses do. The Onyx Coffee Shop, Come to Mama Thriftstore, Purple Circle hair salon, the artsy plastic emporium Plastica, Exene Cervenka's tchotchke shop You've Got Bad Taste, Circus of Books, Mondo Video, Mondorama clothing, Virginia's midcentury modern furniture store—all of them are regular listeners.

The only reason I was at Body Builders Gym was to get in shape for my *Jane* story. I'm applying to be a *Playboy* centerfold and need to send in pictures. I invite my friend Stephanie, the station's unofficial photographer, to shoot them. We are taking

them on a Thursday afternoon in my bedroom at KBLT headquarters, when the light is right and the station is in full swing.
DJ Calamari, host of *Deep Fried*, is on the air, spinning his usual
mix of shit-hot jazz, electronic lounge, funk, and rare soul
when the makeup artist arrives to weed my eyebrows and doll
me up like Brigitte Bardot with fake lashes, baby-blue eye
shadow, and blow-job lipstick. A platinum-blond wig, Calvin
Klein panty set, and turquoise hooker heels complete the look.

The door to my bedroom is closed, so DJ Calamari doesn't
know what's going on. Nevertheless, he's playing the perfect
soundtrack for what I'm doing—the "oh baby" baritone of
Barry White and "sex machine" James Brown. At least he is for
a little while. When he shifts gears and begins playing "Jesus
Christ Superstar" and the Osmonds, I start to lose my mojo. It's
hard enough for me to pout into the camera without cracking
up. When the music is cheese and my wig is slipping, it's almost
impossible.

Stephanie does her best with what's she got—a thirty-one-
year-old quasi-tan body with back fat and a beer belly topped
with cotton-candy hair and an alligator smile. She kills five rolls
of Fuji Gold as I stick out my flat ass, roll around the bed, and
otherwise make a fool of myself. What I do for *Jane*.

While I'm waiting to learn the fate of my *Playboy* photos, I take
off on another *Jane* assignment. The magazine accepted my
pitch on women in the Aryan Nations. My angle: Infiltrate the
group during their annual World Congress and bond with the
girls to figure out why the hell they're involved. It's one of the
stupidest ideas I've ever had.

In my research, I've learned enough about these women to
know they are little more than breeders—passive and sub-

servient to their men. Headstrong and independent, I don't fit the Aryan Nations' MO. I can't go alone. I need to rope some guy into the deal. Someone who looks the part and knows some Nazi lore. I can't ask Jay. He's Jewish. I ask a friend who has a shaved head, but he's not a Nazi. Then there's Don Bolles.

The first time I met Don he was wearing an army-green flight suit, white Snoopy sunglasses, and a coonskin cap with a baby doll's head sewn into its brow. He had no eyebrows. They'd been shaved off and replaced Fu Manchu–style with black Magic Marker. The former drummer for legendary L.A. punk band the Germs walked into the station with nothing but a women's cosmetic case from the 1950s, which he immediately set on the floor and unlatched, extracting a plastic comb and a stick of deodorant before finding the cassettes onto which he'd prerecorded his show. Its title: *Uncle Don and the Secret Hideout Gang Live from the Great Altar in the Middle of the Earth During the Millennium.* It was unlike anything that had ever aired on KBLT previously. The only way I can describe it is a sort of satanic pedophilia on reverb.

These days Don performs his shows live, bringing in about five crates of vinyl each week. I've noticed, in the records he brings to the station, that Hitler cameos on more than a few of his records. Don also told me that he helps his roommate sell SS paraphernalia at the Rose Bowl flea market.

I ask Don if he's interested in an all-expenses-paid trip to Hayden Lake, Idaho. Of course, he agrees.

During our three-day adventure I learn many things. I learn that if you don't wipe your feet on the Israeli flag that doubles as a doormat to the church, your ass is grass. I learn that if you wear brown shoelaces, you must like Mexicans, and for that, you should be prepared to eat shit. I learn that if I didn't have an FBI file before reporting on the Aryan Nations, I certainly do now. And I learn that if I hadn't brought Don along with me, I would never have gotten the story. He is the perfect chameleon.

I've never been happier to return to L.A., even if, as usual, I come home and my house looks like it's been bombed.

CHAPTER

It's six-thirty on a Tuesday in late July and DJ Zamboni, aka Jay, is at the controls with a couple of friends, spinning the Fela/Fat Possum/Oasis/Hendrix/Swans/Verve extravaganza that is *The Magic Ticket* show. I'm in the kitchen, sweating over the stove cooking red pepper sauce and pasta for dinner, when the phone rings. It's DJ Santo. He just got a call from Mark McNeill, general manager of KSCR—the pirate radio station broadcasting from USC. The FCC was just there. They are headed for KBLT.

DJ Santo gives me Mark's number. I dial it with trembling fingers. Mark tells me a KSCR DJ had been knocking on the door to the station when two FCC inspectors approached him and asked who was in charge. The DJ said he didn't know and gave them the studio phone number: 213–740-KSCR. "KSCR?" one of the men asked. "This isn't KBLT?" That's when Mark raced home to call DJ Santo, the only KBLT DJ he knew, hoping he could relay the message to me before the FCC showed up at my house.

By the time Mark and I spoke, the FCC had been gone from KSCR for thirty minutes. KBLT is six miles north of the USC campus. In ordinary traffic, it's a fifteen-minute trip, but that's presuming you know where you're going. The FCC, as far as I know, doesn't have my address. They still need to track my signal to find it. Thank god it's rush hour. Still, I don't have much time.

I thank Mark profusely and hang up the phone, then run to the studio to let Jay know I'm shutting off the transmitter *right now*. I sprint down the hall to the transmitter and flip off the station. In the studio, over the speakers, Portishead's "Glory Box" turns to hash.

I'm shaking. Shit. The FCC is onto us.

I feel awful that KSCR was our fall guy. The FCC does not have an L.A. location. Its nearest office is in Cerritos, twenty miles southeast of L.A. Driving north from Cerritos toward Los Angeles and triangulating 104.7, the FCC field agents ran into KSCR first. The station was a victim of geography. If the FCC had been based north of the city, in the Valley, KBLT would have been busted, and KSCR would be scot-free, at least for a little while. It might not have been for long now that the FCC's on a rampage and shutting everyone down.

The first person I call is Chris W. I want to get the antenna off the roof immediately in case the FCC agents are driving anywhere in the vicinity of KBLT. The situation is grave enough that I also personally contact each of the DJs to let them know what happened. There are ninety-eight of them. It takes hours. I cross calls with a handful who are trying to tune in the station but can't. A couple of them even show up for their shifts, not knowing I've taken us off the air.

For weeks afterward, I'm flooded with calls—from oblivious wannabe DJs who don't even know we're off the air, from listeners asking if we're on summer vacation, and from KBLTers offering help. There's nothing they can do.

Meanwhile, Chris and I are brainstorming about how to proceed now that we know the FCC's onto us. KSCR and fifteen other stations in South Florida have been busted in one week as part of an FCC initiative called Operation Gangplank.

I call Larry to let him know about KBLT's close call, but we haven't been working together so much anymore since I decided to involve Chris with the station. In fact, Larry has become quite bitter about being replaced and is getting a little hostile with me. So when it comes to figuring out how to outwit the FCC and stay on the air, I devise the plan with Chris. Chris and I decide the safest option for going back on air is to use something called a studio transmitter link, or STL. It's a two-transmitter system that relays a signal from my house to a second location that rebroadcasts it.

Broadcasting the signal directly from my house, as I've been doing for the past two and three-quarters years, has always been risky but it is doubly so now that Dunifer's lost his case and the FCC's onto us. If the field agents trace the signal to my house, they are legally entitled to confiscate everything associated with the station—not just the transmitter and antenna, but all the studio equipment, the entire music library, my computer. Stuff that's worth at least fifteen thousand dollars.

With the STL, as the new system is called, the field agents will have to track two signals before they can find my house. They need to find the remote signal first, then trace the originating signal back to KBLT. If they mess with the remote uplink, we'll know in time to cut the signal from my house before they can get here.

Chris's friend "Ed Armstrong"—he took his name from the guy who invented FM broadcasting—has been using a remote uplink system for a little more than a year. Ed designs circuit boards for a living and developed the system himself for Radio Limbo, the 12-watt station he runs in Tucson. He sends a signal from his house to a transmitter in the Catalina Mountains, seven miles from his house and a two-hour hike from the nearest service road. The transmitter is solar powered and battery operated. It runs only at night, to let the batteries fully charge during the day and to allow Ed to be on air when the nine-to-five FCC field agents aren't in their office. For five hundred dollars, Ed says he can build one for KBLT. It will take two months.

Now we just need to find a high-rise—one that is close enough to catch the low-power signal we'll be sending from my house and tall enough that our broadcast won't be blocked by nearby buildings. We'd prefer a low-security structure, one where the guards are asleep at their monitors, but we'll take any building where the roof is not only accessible to deviants such as ourselves but home to other antennas so KBLT's won't seem out-of-place.

Over the next several weeks, Chris and I scout the city looking for buildings. Chris spots a chunky white one at the corner of Wilshire and Vermont that's tall enough, but it's boarded up. There's no way to get to its roof. He thinks the Equitable Building on Wilshire also looks promising, but the security seems too tight and its roof is too clean. KBLT's would be the only antenna up there. I case the Public Storage building at the corner of Beverly and Virgil, circling around it several times before stopping to request a tour. If there's an available cube of space close to a window on a high enough floor, maybe we could rent it and send the signal from there, but there isn't. It's just as well. The place has more guards than San Quentin.

One of Chris's friends suggests the high-rise at the corner of Sunset Boulevard and Vine, smack-dab in the center of Hollywood. His friend is a cook at the 360 Restaurant & Lounge, a hoity-toity restaurant that makes up for its mediocre food with a stellar view, thus the name. The restaurant is in the building's penthouse suite on the twenty-second floor, and a lot of the people who work there sneak smokes on the roof when they're on break.

Chris and I check it out. There's a security guard, but he barely looks at us when we walk through the glass doors. I don't know why these places even bother with the rent-a-cops. Chris and I step into the elevator and ride it as high as it goes, walking the final flight to the roof. The door that leads outside is closed. There's a sticker threatening to sound an alarm if we dare touch it, but Chris pushes it open anyway. The alarm doesn't sound— just like his friend said. We step outside and onto the white gravel. It is littered with cigarette butts and empty beer bottles. Just like home.

The roof has two tiers. Its elevated center section, accessible by a vertical, fire-escape-type ladder, is an antenna farm. About half a dozen radio and television antennas are there, most of them sending signals to Mount Wilson, a peak about thirty miles north of the city, where they are rebroadcast at full strength. We climb the ladder into a gale-force wind and survey the landscape. The view is incredible. It's a clear day, and we really can see forever—north to the Hollywood sign, east to the Griffith Park Observatory, south toward downtown, west to the Pacific. KBLT could broadcast to the entire city with this height. I don't know why I never considered this before. Probably because Larry, who likes to play things safe, never mentioned it as a possibility. It took Chris's involvement and a crisis to prompt the upgrade.

A couple of vacant posts look promising as anchors for our gear, but we still need electricity. All the other stations have their own secured power boxes. We keep looking but find nothing.

We duck back out of the wind and down the ladder to the roof's lower section. There's a single electrical outlet, but it's being used. Chris says he can build a breakout box that will allow us to hack into it without interrupting power to the other device that's plugged into it. Looks like this is the building.

As nerve-racking as it is to have the station down, it is, in a way, a relief. For once, I have my house to myself. Jay and I can sleep there for a change. I can run around naked if I want and talk on the phone without feeling eavesdropped upon. I can do all the stuff normal people do when they're alone in their homes. The station is never too far from my mind, and the phone never stops ringing about it, but at least I can live more or less like a regular person.

I spend most of the summer dodging novice bikers in my motorcycle classes and writing articles for *The Source, Marie Claire, Jane,* and the *L.A. Times,* the latter of which calls to ask me to apply for an assignment editor position that will be opening up in the fall. I am bouncing-off-the-walls ecstatic, even if there's no chance in hell I'll be hired. I've been writing for the paper less than two years, but my editor is asking for my résumé, so I update it and send it in. KBLT is probably my greatest achievement in life, but I don't mention it.

* * *

A call comes in on the voice mail. It's from a stranger. What's new? It's the organizer of Sunset Junction, the yearly Silver Lake street festival with bands, beer, and amusement park rides. More than anything, it's an excuse to show a lot of skin and get smashed during daylight. Does KBLT want a booth for the fair in late August?

Sounds like a good idea to me, but I like to run these things by the DJs and get their take. Jay doesn't like the idea because he thinks it's too high profile and too hot, but almost everyone else agrees it's a great opportunity to do a little public outreach, pocket some coin for the new transmitter, and have fun. We're on. The plan: The KBLT booth will be a sandwich stand, T-shirt/sticker/poster shop, and DJ showcase, even though we won't be broadcasting.

I recruit some of the DJs to set up and break down our booth and assign others shifts for deejaying, sandwich making, and sales. I ask Camille if she'll design the space. She's in, as long as she can also use it to sell her own comics and T-shirts. It's a deal.

Camille thinks even bigger than I do. Her plan includes larger-than-life-sized plywood paintings of her KBLT characters spinning records. They have cut-out faces and hand holes so we can take pictures of people posing as cartoon DJs. She also plans to make two enormous papier-mâché heads that will spin above the tent. *And* she wants to hand build two desks from scratch. She gives herself a week to do this.

Three days before the Junction, the parking lot behind my house is a full-on construction zone. Sheets of plywood are propped up against the house, waiting to be jigsawed by Camille in her safety goggles and gloves, while Chris hammers together two-by-fours, and my asshole neighbor Steve skips rope just inches away to get on their nerves.

Two days out, we are plastering over the center labels of crap records with KBLT stickers. The modified vinyl is part of Camille's master plan for the two desks—one of which will be used as our sales counter, the other as the DJ console. Chris and Camille have built the base cabinets and are slopping resin on their tops to secure the vinyl underneath. The fumes could kill a rhinoceros. The sun is so hot it doesn't just dry the resin but buckle it. I keep Chris and Camille cool with Slurpees from the 7-Eleven around the corner.

The day before the street fair is my birthday, but I won't be celebrating, just working to help get everything finished. I told Jay yesterday that I'm pretending I don't turn thirty-two today. He pretends along with me and doesn't even wish me a happy birthday.

The plywood cutouts are only partially painted. The plaster heads are still drying. And the two eight-foot-tall signs announcing "Sandwiches, Tees, Tanks, Stickers, and Grooves!" and "Posters, Comics, Ice Cold Water, and Photos!" have only been stenciled. Camille recruits her mother, a muralist, to finish hand lettering them.

Eddie Muñoz, guitarist for the Plimsouls, has the largest collection of psychedelic print shirts this side of England and a sixties garage-rock show on KBLT that would blow the skinny little tie off the Kinks' Ray Davies. Even better, Eddie has a Costco card. He picks me up in his car and we drop about a hundred dollars buying bulk packs of bacon, a jumbo jar of mayo, twelve loaves of bread, cases of beer, and enough snack-sized bags of chips to keep an eight-year-old fat and happy for an entire school year.

Camille's mom takes advantage of my absence to buy me a birthday cake, which we slice and serve in my living room, under the shadow of enormous plywood walls and hulking papier-mâché heads. Chris, Camille, and her mom are there. So is Larry, who has been in town two days and done nothing but sulk and lumber around the house. Camille has been calling him The Blob behind his back. About half the cake is left after each of us has had a piece. Larry pushes it onto his plate and goes to town.

It's after midnight when everyone clears out, but not for long. They're back at seven the next morning to load up one of the DJ's pickup trucks and head to the exhibit area. Our booth is only a block from my house, but we need to truck everything in via Santa Monica Boulevard, which is backed up and bumper to bumper—a hipster caravan of cars and vans overflowing with shirtless boys and next-to-naked girls. It's like *The Grapes of Wrath* with tats.

The line starts moving at eight, when the security guards, clipboards in hand and whistles around their necks, let us pass through the chain-link gate. We find our tent, marked with an index card duct taped to the ground. I glance at the space next to ours. There's a ten-foot stack of speakers and a banner: GROOVE RADIO 103.1.

What the fuck. They intentionally put us next door to another radio station? I know the Junction's organizers try to cluster similar services, but this is ridiculous. Groove Radio couldn't be a better metaphor for what KBLT is fighting on the FM dial.

I look across the aisle. There's an empty space. I glance left, then right. An Asian couple, gas grill in tow, appears to be heading for it. I rip our card from the ground, and make like Flo Jo, dashing across the concrete to swap cards.

We get everything together in two hours and the fair begins. Our booth is orange. On its front are the letters *BLT* in glitter. We don't use the *K* because I don't want to attract unwanted attention. So what if the station's been on the CBS national news and in various magazines? After our close call with the FCC, I'm being extra cautious. My logic: The station's fans will figure it out. Everyone else will just think we're selling sandwiches.

A few people stop and give up their pocket change for a sticker, but the BLT stand is mostly slow 'til noon, at which point the only in-tent exchange is excuse-me-pardon-me as the DJs try to move around one another. The booths are only ten square feet and there are five people on shift at all times—two to work the sales counter, two to assemble sandwiches, and the DJ. It is worse than McDonald's on a Monday lunch break, with people rummaging through bales of tank tops, DJs thumbing through their record crates, and short-order cooks wielding knives—all of them ducking beneath the dozen or so shirts hanging from our clothesline display.

A sandwich is three dollars. Give us double that amount, and we'll throw in chips, bottled water, and the Hostess product of your choice. T-shirts and tanks are fifteen dollars. Posters from our two benefits are ten dollars. Stickers, a buck. At these prices, no one's going to get rich, least of all KBLT. We'll be lucky to make our money back, especially since all the DJs are gobbling down the BLTs and gulping beer from the Igloo. Oh well.

With the herd of friends and curiosity-seekers who've come to my house, I feel like I've met everyone who listens, but there are a lot of unfamiliar faces stopping by to see what we're about and to confirm the juicy rumor that the FCC's on our tail. They want to hear the drama firsthand.

Watt is one of the first DJs to go on. It's one o'clock when he snaps a Temptations platter·on the turntable. It is so hot the record warps midsong. If he doesn't get his crate of records out of the sun in the next sixty seconds, his highly collectible vinyl will curl like kettle chips and be just as worthless. We rig a canopy of towels over the turntables, which seems to do the trick.

The speakers are harder to deal with. Watt is spinning Billie Holiday, but all I can hear is Groove Radio, which has been blasting indistinguishable monobeat dance tracks for the past three hours. I look at our speakers. They're the six-inch cubes we use in the studio. I look at Groove. They've got towers of woofers and tweeters, which are doubling as dance boxes for gym victims in hot pants and work boots who are pelvic thrusting to the beat.

I'm halfway up the street to my house to grab the pair of slightly larger speakers that go with my bedroom stereo when a guy calls out to me from his car. He looks familiar, but I don't know his name. "Tony," he says, extending a hand. Unknown to me, he was deejaying at my house nearly every week before we shut down. I fill Tony in on the sound disaster down below and he offers to loan us his gear, the pro kind. I'm not sure how long it will take him, so I pick up my home speakers to tide us over.

Just as I'm hooking them up, DJ Santo arrives at the booth. His record bag is brimming with white-label vinyl, and he's ready to pick up where Watt's about to leave off. I know he also has professional speakers from spinning gigs around town. I beg him to go home and get them, and—while he's there—how about his mixer too, since KBLT's is acting up.

With Tony's and DJ Santo's sets of pro speakers, we can at least hear KBLT inside the KBLT booth. Groove is still louder, but the Junction's security people, for some reason, are asking us to turn down our music and threatening to kick us out. One of the times they approach us, DJ Reverend Mo is spinning a record by lounge icon Martin Denny. We're playing easy listening, and they want us to keep quiet? Unbelievable. We keep the volume cranked but angle our speakers away from Groove, keeping a lookout for security so we can turn down the knob on a second's notice.

I suspect the people at Groove sicced the security on us, though it's possible the festival's organizers are after us because we're causing too much of a scene. A girl is dressed like one of Camille's KBLT characters—a sort of demented French Ed Grimley, with a cowlick and red-and-white striped shirt. She has pulled one of the arrow signs off our booth and is doing a *Saturday Night Fever*. Some of the girl DJs decide to join her, choreographing a routine that devolves into a mini dance party when onlookers join in.

I've arranged for some of the more danceable DJs to spin after dark. The crowd, drunk from downing plastic cups of Bud, gets larger and wilder as the night wears on. One guy, in a headlamp and shorts, clears a circle around him with his crazy dancing. KBLT is a freak magnet.

The fair shuts down for the night, and I sleep at my house because it's just a half block away from the KBLT booth. The fan in my room is worthless. By eight the next morning, I'm already sweating. Even worse, I'm hung over. My tongue is dry as cloth, my skin pure slime. I force myself to the kitchen to make coffee for the troops who are supposed to help me move everything back to the booth, but no one shows. I grab the CD player that was ditched just inside the front door when we hauled everything up to my house last night and head outside. A grocery cart is parked on the sidewalk. Should I? Could I? I *really* don't want to be seen doing this, but I load it with speakers and ride it down the hill to our booth, then return for another dozen loads. The DJs arrive about fifteen minutes before the fair's supposed to open.

CHAPTER 16

It's early October, a week before KBLT is scheduled to go back on the air, and I'm scheduled for an interview with the *Los Angeles Times*. It's a three-hour interrogation with six separate interviews, beginning with the editor of the paper and continuing all the way down the chain of command.

It's been a long time since I've had to dress better than a street urchin, but a suit is too great a leap. I settle on a conservative blouse and pants, even though my hair still screams punk rock. It's so black it's blue. Oh well. There's nothing I can do about it now.

The Reverend Mo is at the station as I'm doctoring my makeup and gussying up. He's organizing the library for the DJs, who will begin swarming my apartment again next week. Hearing the click of heels on hardwood, he looks up. "Wow, Paige. I almost didn't recognize you."

"It's a leap, I know."

"Well, you look very nice," he says, nodding his head in approval.

"You think? I just hope I don't trip in these things." I hold out a foot bound to a four-inch black sandal that is dangerously close to a hooker heel.

"You look good—straighter than I've ever seen you—but good." He pauses, looking me up and down. " It's none of my business, I know, but what's the occasion?"

I tell him I have an interview with the *L.A. Times.* He seems surprised. Like most of the DJs, he knows I write for *Jane,* but he has no clue I also write for the paper.

I excuse myself so I can look in the mirror a thirty-second time, making sure my hair isn't acting all Bozo and that there's no toilet paper stuck to my shoe, then run out the door to Jay's car, which I've borrowed for the day. I doubt that leather, heels, and a helmet would go over too well.

I'm thrilled to be interviewing at the *Times,* but it makes me feel like a traitor to the radio station. The only time the DJs have ever heard me talk about the paper was when the unauthorized article on KBLT came out a little more than a year ago, and I was spewing pure venom. Getting the job would be an enormous coup for me as a journalist, but I have no idea how I'd simultaneously edit for the newspaper and run a full-time radio station. I'll cross that bridge when I come to it.

The weekend before we plan to go back on air, I hold a meeting to explain the new system to the DJs and to institute some new rules. The DJs are no longer allowed to swear, to say the word *pirate,* or to mention anything that might give away our location. As much as I'd like to maintain a free-speech ethos, there's too much at stake. I don't want anyone saying something that could (1) piss someone off and prompt them turn us in or (2) give the FCC any clues about what and where we are. There's a little grumbling among the DJs, but they understand the station's had one close call and probably won't be so lucky next time.

Chris drives to Tucson to pick up the equipment. Ed could mail it, but some of the pieces are weird-shaped and fragile. Besides, Chris needs to learn how to set it up because the new gear is different from what we had been using.

The new setup is sophisticated but simple. We transmit a 2-watt signal from KBLT on the stereo audio carrier of an unused UHF channel. That signal is caught by the receiver we'll be installing on the 360 building via an antenna tuned to the same frequency. It is then remodulated to 104.7 FM and transmitted out of another antenna at 8 watts. We're dropping our power to about a fifth of what it was with the old system because if we continue to broadcast at 40 we'll be interfering with the nearest licensed station on the same frequency, sixty miles north in Oxnard.

Chris sets up the equipment to broadcast from my house by himself, but getting the receiving and retransmission gear in place at Sunset and Vine is more complicated. There's too much equipment for Chris and me to carry alone, so we recruit Brian—host of a Thursday morning down-tempo country and western show called *The Three Sixes*—six strings, six packs, and six guns. Brian has helped with a couple of other station projects—a failed attempt to lift the antenna (again), and the installation of shelves to house our ever-increasing CD collection.

We pick a day midweek, late in the afternoon, and we meet at my house to gather tools, get the equipment together, and coordinate our game plan before trotting off to the high-rise at the intersection of Sunset and Vine in Hollywood—a location we're keeping top secret. We're not even telling the DJs.

Antenna mast: check.

Receiving antenna: check.

Transmitting antenna: check.

Power supply, studio uplink unit, and transmitter: check.

Electrical cable and hardware: check.

We load everything into the back of Chris's station wagon, leaving a CD playing on repeat at the station so we can verify we're broadcasting once everything's in place. In the car, we do nothing but run through the plan ad nauseum. All of us will carry the equipment to the roof. Brian and I will attach the main antenna to the mast while Chris connects the power cable. All three of us will erect the mast together, attach it to the power grid on the building, and install the receiving antenna. Chris will set up the transmitter box, hook it up to the antenna, and flip the switch. Then we will run away.

We park in a strip mall kitty-corner from the building and carry our equipment into the lobby. We are all wearing blue, dressed in whatever version of a service worker's uniform we could derive from our closets. If anyone asks, we're delivering utensils and cookware for the restaurant. That doesn't explain the ten-foot pole and two antennas, but it's our best excuse.

No one else is in the lobby but the security guard. The elevator is there and waiting. We pile in, but the mast—the ten-foot pole—doesn't fit.

The security guard gets up from his desk and saunters over. "That ain't gonna work," he says, holding the door. "You're gonna have to walk that one up. What floor are y'all headed to?"

"Twenty-two." Chris doesn't say anything else, and the guard doesn't ask.

He just shakes his head. "I'm sure glad I ain't you."

Chris volunteers to be stair master, though any one of us

could have. We are all so jacked on adrenaline we could sprint to the viewing deck of the Sears Tower and go skydiving afterward.

Brian and I ride the elevator to the twenty-second floor, then walk the last flight up with our boxes. Waiting just inside the door to the roof, we hear Chris's footsteps. "Six floors," he counts down. "Five."

He rests a few minutes once he gets all the way up, then we press open the door. We want to make this as fast as possible, on the odd chance someone will catch us in the act, but climbing to the elevated portion of the roof is tricky with all this gear. The ladder is thirty feet high and 100 percent vertical. We should have brought backpacks, but no one thought of that, so we're left balancing the boxes between our knees and elbows while we climb.

We're all scrambling with our tasks when we hear the door slam. Was someone just up here? We try to move faster, but the Chicago-style wind makes it difficult. We are the scene from Iwo Jima as we raise the mast.

Thirty minutes later, and everything's good to go. Chris pulls the Walkman radio from one of the boxes and tunes to 104.7. Mercury Rev's *Deserter's Songs* is coming through loud and clear.

We pack up our stuff and hightail it out of the building and into the car. Driving east down Sunset Boulevard toward my house, the signal is strong and solid. I feel invincible.

Starting with Jay's show at six, which he now calls *Radio Free Albemuth,* the DJs are coming in for their regular shifts. Eddie Muñoz rocks out with psychedelic pop and garage rock at eight. Laurel does her punk and prog thing at ten. And so it goes, around the clock without a break. KBLT is back in action.

The mood at the station has never been more electric. Part

of it is everyone's enthusiasm to be back on the air, part of it is
reveling in the glory of having outwitted the FCC, and part of it
is that our signal is reaching farther than it has ever reached before. DJs are calling in with status reports from as far south as
Irvine and all the way out to the ocean, twenty-five miles west,
saying the station is coming in loud and clear. Finally, we have
Santa Monica.

It would be nice to believe our listeners have been tuning to
104.7 occasionally in the three months we've been down to see
if we're back on the air, but I suspect many of them gave up on
us sometime after the Sunset Junction street fair. At that point,
we were telling them we'd be up and running in early September. It's October 13.

To announce our return, one of the DJs suggests we hijack a
couple of billboards, plastering over them with our own image
and logo. Renegade advertising for our covert operation. I love
it. Camille designs the poster: The KBLT girl is riding a pig, a
radio tower is on the horizon. Neither the station nor its call
letters are mentioned, but anyone familiar with KBLT will recognize our mascot and understand the slogan: WAKE UP AND
HEAR THE BACON.

It is still dark when, at five in the morning a couple of days
after we flipped the switch, I'm driving around the neighborhood with Brian (who had just helped install the new equipment) and another DJ named Carolyn. We're looking for
billboards that are (1) low enough to the ground that we can
get to them without falling to our deaths and (2) dark enough
that any cops cruising the neighborhood won't see our silhou-
ettes and arrest us. We're all dressed to look like billboard in-
stallers, in caps and painter's pants, but we're the Three Stooges

when we attempt to cancel out a Spanish Cheerios ad on Sunset Boulevard and a Colgate ad in the parking lot of the House of Pies diner about a mile away. Climbing up to the signs, we take turns clinging to the billboards for our lives and flinging wheat gluten to paste down the posters.

Either Steve Malkmus has discovered the distortion pedal or KBLT just lost its signal. The lilting bounce of Pavement's "Westie Can Drum" has just switched to fuzz. Ordinarily, I'd have noticed something so major, but today I'm not listening very carefully. I'm at my computer, finishing a story for *Jane* about the all-woman chain gang I worked on last week in Phoenix.

The phone rings. I push back my chair and pick up.

"Hey, Paige." It's Chris. "Is the station on right now?"

"I think so," I say, not really knowing. I'm still thinking about the tumbleweed I was forced to roll while wearing black-and-white-striped PJs. "Why? What's up?"

"Are you sure it's on?" Chris's demeanor is normally a lesson in calm, but today there's a hint of panic in his voice.

"Hang on." I lean through the kitchen door, listening for sounds from the studio. I hear something but don't know if it's music. "I can't tell," I say, wanting to get off the phone and back to my story. "You're not picking it up at your house?"

"No." He is definitely panicked. "Something funny just happened. The station clicked on and off a few times. Can you double-check we're broadcasting?"

Chris's worry is contagious. Suddenly I'm snapped out of storyland and focused on KBLT. The station's been on the air for two weeks, and we haven't had any problems. There's no reason why Chris shouldn't be hearing it in Echo Park, just a few miles away.

"Sure," I tell him. "Give me just a sec." I let the receiver dangle from its cord and speed walk to the studio. DJ Kari Kaos is at the controls, headphones around her neck. She is flipping switches and looks confused.

"What happened?" she asks me. "There's no sound."

Shit. I run to the kitchen, fumbling for the phone. "You're right. We are down."

The station has always been rife with technical problems. It hums. It distorts. It's burned up. It's lost the frequency. You name it, we've experienced it, but something seems different this time.

Chris asks me to check the power meter—a gizmo that registers how many watts we're transmitting. The needle is registering at 2, just like it's supposed to. He asks if the uplink box is plugged in and turned on. It is. If everything is working from our end, something must be wrong with the remote transmitter on the high-rise at Sunset and Vine.

Chris is on my roof within twenty minutes, checking the uplink antenna. It's still connected and transmitting at the right frequency. Everything's kosher. He confirms the problem is not at my house but in Hollywood.

My motorcycle is faster and easier to park than Chris's wagon, so we take the bike, parking between cars on Sunset and rushing up the elevator but slowing down as we approach the roof's door. What if someone's waiting for us? Are we walking into a trap?

We have no idea. We don't even know what's wrong, only that KBLT's off the air. It could just be a technical problem. Maybe a wayward seagull took a dump on our rig and it oozed into the transmitter. Maybe the wind bent our antenna. Pushing the door open slowly, we step outside. The gravel crunches beneath our feet. There's no hiding the fact that we're here if anyone is waiting for us, but we don't see a soul. Perhaps they're on the elevated section of the roof. Lurking.

Ever chivalrous, Chris scales the ladder first. Cautiously, he pokes his head over the final rung, scanning for an ambush. It's clear. He steps off the ladder and onto the roof, motioning for me to do the same. He's looking into the transmitter box when I walk up behind him. He turns toward me.

"The power switch is turned off."

Someone turned off the station, and whoever it was knows radio equipment. Instead of snipping the cable, they unscrewed the cover of the transmitter box, found the power switch inside, flipped it off, and put the box back together.

Standing on the roof, with the sun singeing our scalps and wind swirling around us, we discuss the possibilities. It could be that one of the legit stations in the building saw our equipment, knew it didn't belong, and thought they'd have some fun. Or it might be another pirate trying to take out the competition. Maybe it's Larry, since the two of us aren't talking right now. I just hope it isn't the FCC.

Regardless of who did it, we need a plan. We could go back to the house for tools and take the equipment down, but we don't want to. In the short time we've been back on the air, the response has been amazing. The DJs have never had so many calls. It's Thursday. If it is the FCC or some other working stiff, they've only got one more day before the weekend. What are the chances they'll do it again Friday? We turn the station back on. We'll move it Saturday, though we don't know where.

Chris and I scan the roofs of nearby buildings. The high-rise directly across the street might be a possibility. We leave the 360 and walk into the building on the other side of Sunset. The lobby is far more posh. The security guard is master of an entire island. He is staring at us. We're probably the only people who've come to the building today wearing anything other than a suit.

Feeling his eyes, and looking for an excuse to head for the elevator bank, we check the directory. We decide we are visitors to the accounting office on the nineteenth floor, the highest one we can find. From there, it's only a few flights up to the roof. Only this one doesn't have any other antennas or dishes. There's no place for ours to hide. We go back downstairs and out to my bike. We'll need to think about this one overnight.

CHAPTER 17

It's the day before Halloween, and Kari Kaos is playing Marilyn Manson when Chris and I walk in the door. Both of us suspect KBLT won't be on the air much longer, but the DJs continue to come to the station as if nothing happened. DJ Calamari stops in to spin T. Rex and Queen. DJ Mario plays the Carpenters and Gil Scott-Heron. *Three Straight Guys in a Closet*, as they call their show, spin Primus, the Deftones, and Rage Against the Machine.

At 10 A.M. Friday, it's the *Morning Mind Melt*—a show where krautrock meets dub meets the avant garde. This morning, Eddie, who hosts the show, is exploring what he calls "the dark side of music," spinning Chick Corea, Kenny G., and other saccharine jazz acts with every available effect on the mixer. The microphone is heavy with reverb, and he's using listeners' messages as samples, which he slows down and drops into the music at whim. He's on the air less than an hour when another effect kicks in: static.

"Paige?" Eddie calls from the studio. "There's something wrong with the radio station."

He's barely finished his sentence, and already I'm hyperventilating. I run to the studio. Just like yesterday, static is blaring from the speakers.

"Is everything okay?" Eddie asks.

"Not really." I feel sick, and I'm shaking. "This might be it."

"What? The FCC?"

"I don't know, but someone's definitely fucking with us," I say, leaving the studio for the living room.

Eddie follows me.

"Can you keep deejaying even though the station's down?" I ask while I zip on my boots.

"Whatever you need, Paige."

I grab my helmet and head for the door. I run to my bike and fire up the engine, leaving the choke on as I wheel out of my driveway toward the intersection of Sanborn and Sunset. The light's red. I nervously tap my heel on the pavement, racing onto Sunset the second the light goes green. My adrenaline level is no match for the traffic, which is slow and sticky. I scream at the cars to get out of my way, but they're like roadblocks.

Weaving my way through the cars, I pass the Akbar, a favorite DJ hangout, and Circuit City, where I've dropped several paychecks. I race by Orchard Supply Hardware, where Larry and I spent so much time collecting parts for the station, and pass the club where Mazzy Star played KBLT's first benefit concert. This can't be happening. KBLT's been back on the air only two weeks. The new transmitter system can't be that easy to detect.

I continue down Sunset, training my eyes on the high-rise in the distance. It's so far away the building looks like a miniature souvenir. I refocus on the street, angling my bike around potholes while I work on doubling the speed limit. Twisting the throttle, I pass people in the center lane, praying no one will make any sudden moves that will send me flying.

It's hot. My head is sweating into the lining of my helmet, and I can feel the heat from the exhaust pipes on my thighs. I'm getting close enough to the building at Sunset and Vine that I can see something moving on its roof—*people* moving. My muscles turn to jelly. I can barely keep my bike upright, but I'm compelled forward by my need to know who those people are. They are on the northeast side of the building, fiddling with the KBLT antenna.

I park on Sunset across from the building and run into the lobby. The elevators are on other floors. Everything seems to be working in slow motion today. I watch the lighted numbers on one of them tick down to the lobby and rush the doors once they open. I press Close Door before anyone else can get in and take it to the top.

The hallway is quiet when I step out onto the carpet. I open the door to the stairwell. Voices echo on the steps above me. My stomach flips. No one takes the stairs in L.A., just like nobody walks. I should turn around now and leave before something horrible happens, but I need to know who's messing with KBLT.

I creep up one flight of steps without seeing anyone, then another. The voices are getting louder. Coming up on the landing that leads to the roof's door, there's a stubby man with buzz-cut, silver-gray hair, and ruddy skin. He turns around halfway and looks at me. Cocking his head and pointing a thumb in the general direction of the roof, he asks, "Are you heading up there?"

"I'd like to talk to you about that," I blurt out, immediately regretting my words.

He smirks and reaches for his wallet, pulling out a business card and holding it in front of my face. The lettering on the card is so small that it takes a few seconds to register, but then I see it: FEDERAL COMMUNICATIONS COMMISSION. I want to heave.

"Is that equipment up there yours?" he asks.

I nod even before I realize what I'm doing.

"I'll tell you what. You've got two options," he tells me. "You can sign the equipment over to me or take a fine of ten thousand dollars."

The equipment cost five hundred. I tell him he can keep it.

"Where was your operation based anyway?"

Without hesitation I tell him: "Silver Lake." For some reason, I'm telling this guy everything. I'm not behaving like Paige Jarrett—the tough-stuff chick who'll do whatever it takes to keep her station up and running. I'm acting like Sue Carpenter—the responsible and honest midwestern girl I am at my core. I shouldn't be giving such top-secret information to the FCC, but I'm scared. I've never been in trouble with the law before, and these are federal agents.

He leads me up another flight of stairs and into the 360 restaurant, where I meet FCC agent #2. Agent #1 looks like he did time with the military, but this second agent is a hippie lost to the dark side. He's got a long gray ponytail and is wearing a T-shirt and jeans. He also smiles when he sees me. Out of politeness, I smile back. I could kick myself. What the hell am I doing? This isn't going even remotely the way I'd planned. I should be bashing them with my helmet, kicking them both in the balls, and making a fast getaway. Instead I'm standing here hobnobbing.

Agent #1 extracts a piece of paper and hands it to me. It's an acknowledgment that I own the equipment and that I'm officially relinquishing it to the FCC. I sign an illegible "Paige Jarrett" on the line near the bottom and hand it back.

"What does that say?"

"Paige," I say.

"Last name?"

"Jarrett."

He doesn't even ask for my ID.

I ask how he found out about KBLT. He says there was a complaint against the station, but he won't tell me the nature of the complaint or who filed it—that if I want any more information, I'll need to file a Freedom of Information Act request with FCC headquarters in D.C.

"That's it, right? I can leave now?" I ask.

They nod. I yank open the door to the stairwell, bursting into tears the moment it clicks shut behind me. I'm simultaneously sad the station is busted, angry to have been caught, frustrated that I can't do anything about it, scared something else will happen to me, and confused about what to do next.

Driving home, I feel disconnected from my body. My mind is far from the road as I wallow in self-pity and scheme up ways to keep the station alive. Another rig will cost five hundred dollars and take a few months to build. Judging from this last stretch on air, it would probably buy us only another two weeks. And then what? Build yet another rig and move to yet another rooftop? This could be a very expensive game of cat and mouse—one that could ultimately land me in jail.

As much as I love the station, in my heart I know my days as an FM radio Robin Hood are over. I dread telling the DJs, who devoted so much time, energy, talent, and enthusiasm into building KBLT into what it is—or, rather, was. I'm sobbing when I ride up to my house. Eddie is outside talking with Hassan, the jazz DJ whose show follows his.

I pull myself together, park my bike, and give them the bad news: KBLT is over.

EPILOGUE

A few days later, I was four thousand feet above Los Angeles, flying in the helicopter I'd rented so Chris W. and I could scout the city's rooftops for a new transmitter location. What can I say? KBLT was worse than a heroin habit. I couldn't give it up.

I'd just watched my little idea seed an entire community. I'd watched the station evolve from a single person to a round-the-clock operation with hundreds of DJs. I'd watched the station grow from having no listeners to being profiled on the CBS national news. I truly felt I had created something special. Through KBLT, I'd found my strength. When people believed in what I was doing, I started to believe in myself, and that wasn't a feeling I wanted to let go.

So I kept trying to make it work. A few weeks later, on what would have been KBLT's three-year anniversary, we went back on the air, setting up the station the old way and broadcasting directly from my house. It was Thanksgiving weekend. If we were on vacation, we figured the FCC was, too. We were on the air four days. Just like before, the DJs came in all day and throughout the night. But something had changed. Instead of being upbeat and playful, the DJs were gloomy and morose. It just wasn't fun anymore. They knew, as I did, that we couldn't continue like this.

The FCC had already mailed me a notice, telling me I'd be fined ten thousand dollars if I went back on the air. They sent it to Susan Carpenter and mailed it to my house. When I called the agency to find out how to file a Freedom of Information Act request and learn who had turned in KBLT, I'd inadvertently revealed my real name and home address. I didn't know my phone's caller I.D. blocking was voided when you call government agencies. As for who filed the complaint that prompted the FCC to come after us, according to my file, that was the FCC complaining to itself. The Bay Area field office found out about KBLT through an on-line pirate radio fan site and tipped off the office near L.A.

KBLT went dark after Thanksgiving, but I was hopeful the station might rise again. Months before I was busted, the FCC began to consider opening up a new type of radio station for low-power broadcasters called LPFM, for low-power FM radio. The new permits were going to be available to community groups and would allow them to broadcast at 100 watts to a radius of 3.5 miles.

The FCC said the permits would be available even to former radio pirates, as long as they had stayed off the air once they'd been caught. There were indications the LPFMs would not be available in large metropolitan areas like Los Angeles. Supposedly, there wasn't enough room on the dial, even though KBLT and dozens of micro radio operators in other major cities around the country had proven otherwise. Competition for the LPFMs was expected to be fierce. In most cases, a single frequency would be opened in any one market and only one group would get it. Dozens of organizations could be vying for a single spot on the dial, and the permits would be given to groups that had a strong history in the community.

That was KBLT. So what if we'd gone back on the air for a long weekend after the bust? The FCC would never know. There was hope. A couple of the other girl DJs and I organized the Jerky Girls project. For about a month, we dressed up like cigarette girls and canvassed local nightclubs to collect signatures and drum up support for the station in the event that the FCC adopted the new LPFMs, which it did, about two years later.

Kennard, it turned out, was one of the good guys. He may have sicced his enforcement officers on a group of Florida pirates immediately after he took office, but somewhere along the line he had a change of heart. During the three years he was FCC chairman, he was an eyewitness to what the Telecommunications Act of 1996 had really accomplished—putting more and more radio stations into fewer and fewer hands. At the same time, he had watched the micro radio movement escalate to a peak never before seen. Clearly, there was a need for more airwave accessibility.

Shortly after KBLT stopped broadcasting, the FCC began studying the technological feasibility of LPFMs. It conducted an engineering analysis to determine whether the addition of low-power stations to the FM dial would cause interference to those already there and to figure out how much space would need to be left between stations' frequencies to prevent that from happening. The National Association of Broadcasters and the National Lawyers Guild commissioned their own studies on the subject. Not surprisingly, the results of each group's studies dovetailed nicely with their interests.

The NLG, the group that had represented Stephen Dunifer throughout his lawsuit, found no significant interference. The

NAB, the powerful Washington, D.C., broadcasters' lobby that had helped push through the Telecom Act a couple of years earlier, found there was too much interference to allow LPFMs. NPR came to the same conclusion and, in what some critics called an unholy alliance, threw its support behind the NAB.

Surprisingly, the FCC found no technical merit to the NAB's study. Its results synced up with the NLG's, finding there was no objectionable interference between stations, as long as they were spaced on third-adjacent frequencies. In other words, as long as they were three clicks away from the nearest station in either direction on the dial. The nearest frequency to a station on 104.7, for example, couldn't be any closer than 104.1 or 105.3.

The NAB got all hot and bothered by this, of course. Allowing low-power stations on the dial with third-adjacent spacing would mean a couple of thousand additional stations on the dial across the country, and additional stations meant more competition. More competition meant fewer listeners, and fewer listeners meant less advertising, which meant less money. So the NAB did what any self-respecting corporate lobbying group would do: It knocked on select congressmen's doors to have the FCC's study results thrown out. And guess what? That's exactly what happened. Through a perversely titled law called the Radio Broadcasting Preservation Act, Congress decided to allow the new LPFMs, but only if they were even farther away from each other, on fourth-adjacent frequencies, or four clicks away from the next nearest station on the dial. The new standard killed 80 percent of the planned new stations before they even had the chance to go on air.

The only place where there's enough room for fourth adjacencies is in minuscule towns that really are forty watts from nowhere, so the approval of LPFM was a bittersweet victory.

On the upside, there will be about a thousand more independently operated community stations on the air around the country once the FCC finishes sorting out applications and doling out construction permits. (As of this writing, four years after LPFMs were approved, only half of them have gotten up and running.) On the downside, the new LPFMs don't do anything to solve the airwave inaccessibility problem in even moderate-size cities, let alone huge metro areas like Los Angeles, New York, Chicago, Philadelphia, and Phoenix.

As long as that's the case, there will always be people who take to the airwaves illegally. In fact, many of the illegal stations that were on the air at the same time as KBLT are still operating, even those that were caught by the FCC after Dunifer lost his case.

Stephen Dunifer may have been barred from broadcasting and from aiding anyone else who wanted to do so illegally, but that didn't stop Berkeley Liberation Radio from picking up the cause where he left off. Moments after Judge Wilken's ruling, BLR jumped on Free Radio Berkeley's frequency. They broadcast for nearly four years without any trouble from the FCC, until March 2003, at which time the station was raided by a team of U.S. marshals, Oakland cops, and FCC officials who took the station's door off its hinges and pointed three .38-caliber pistols at the DJ's head. The station went off the air briefly but has since gone back on the air.

San Francisco Liberation Radio also went dark immediately after Judge Wilken's ruling but returned to the airwaves seven months later, having applied for an FCC license without ever receiving a response. That was February 1999. One month later, the FCC went to Richard Edmondson's door and handed him a threatening letter, which he chose to ignore. The station eventually changed hands and locations, upping its power to 100 watts and moving to higher ground that allowed the signal to reach not only San Francisco but Berkeley and Oakland. SFLR operated in the clear for a couple years until October 2003, when ten federal agents, ten San Francisco police officers, and five FCC agents raided its broadcast studio with a battering ram and firearms, confiscating all the station's equipment.

Radio Limbo, the Tucson station run by equipment supplier Ed Armstrong, was raided just a couple of days after KBLT, despite its stealth location two miles deep in a national forest. One morning Ed was listening to a DJ play a few wacky tunes. That afternoon it was all gone. When Ed hiked into the woods, nothing was left but a few strands of baling wire and a few boot prints. In late 2002, Limbo went back on the air. It hasn't been visited by the FCC.

KSCR, the Los Angeles micro radio station KBLT inadvertently busted, moved their operation on-line to kscrradio.com. In January 2003, they began simulcasting on the AM band. Station manager, and KBLT savior, Mark McNeill went on to co-found the cutting-edge online radio station Dublab.

As for KBLT, a number of its DJs joined a politically oriented on-line station called Kill Radio. In October 2002, some unknown person began simulcasting on KBLT's former FM frequency. Who's pirating the signal? The DJs say they don't know. Even if they do, they're not telling, especially not me, since I'm now part of the media establishment.

One has to wonder why the FCC is letting the micro radio movement continue to quietly percolate and why it's so inconsistent in enforcing its own policies. Maybe they're tired of fighting a lost cause. As part of the Radio Broadcasting Preservation Act, the FCC was allowed to test more closely spaced LPFMs in nine markets, with Congress having the final say over whether those tests were successful enough to allow more stations. In the summer of 2003, the results were published in a report that recommended the present third-adjacent channel protections be waived because "perceptible interference caused during the tests . . . occurred too seldom." Whether this study will result in more LPFMs is uncertain. Actually, it is unlikely. Passing legislation is extraordinarily difficult; reversing it is almost impossible. Set into motion with the 1996 Telecom Act, radio deregulation is already entrenched.

As of this writing, Michael Powell is chairing the FCC with a decidedly free-market bent. In June 2003, he relaxed restrictions on cross-ownership of newspapers and television stations. That prompted a wave of public outcry over the loss of media localism and, as a result, Powell's announcement three months later of an initiative to promote local content in radio and television. Though Powell said he's unsure whether diminishing local content is really a result of fewer broadcast owners, he did nevertheless vow to look into new rules that would make broadcasters more responsive to the communities in which they operate and to accelerate the licensing of LPFMs. The Prometheus Radio Project, a Philadelphia-based community-radio advocacy group, has since sued the FCC over the new rules, successfully persuading a federal appeals court to issue a stay on their implementation.

In 2005, the whole issue of media diversity will again come up for discussion when the FCC reviews its rules, as it does every other year. If Powell remains in office, he could very well lift the remaining ownership caps that currently restrict a single broadcast company from operating more than eight stations in a single market.

If that happens, the micro radio movement might get more active, but it will probably never reach the fever pitch it did in the mid-1990s, when Dunifer was waging his First Amendment war with the FCC, especially now that LPFMs have been approved and because there are other opportunities for people to express themselves, like the Web.

While it seems like it would be relatively safe to join the ranks of my fellow micro radio operators and return KBLT to the FM dial, I've chosen to remain off the air. Remember that editor position I interviewed for at the *Los Angeles Times*? By some bizarre twist of fate, the newspaper called to offer me the job the same day I raced my motorcycle to the KBLT transmitter and inadvertently turned myself in to the FCC. I was on the job three weeks later and have been there ever since.

There are times I desperately miss KBLT—when I unexpectedly run into a DJ on the street or when I'm listening to some really great music and want to spread the word. But I'm satisfied to know I played a part in changing the rules. I'm satisfied to have built what may have been the largest and most popular pirate radio operation in L.A., and satisfied to have become, in a way, a sort of Stephen Dunifer. Through KBLT, I learned that I had the courage to work outside the system to make an unusual project succeed. I learned that with effort and determination I could create big things, and that those big things could positively affect others' lives as well as my own.

I know all this sounds hokey. It's the classic American dream story, even if it does have an illegal twist. I did break the law, after all, and break it in a pretty big way. But sometimes, just sometimes, that's a good thing.

ACKNOWLEDGMENTS

KBLT may have started as my idea, but it was not my success. That credit goes to the collective will of all the people who got involved because they loved the idea and cared enough to make it work.

First, I'd like to thank Stephen Dunifer and Free Radio Berkeley for having the balls to go up against the FCC and for creating the legal loophole that allowed operators like myself to go on the air; Peter Franck for mentioning his case to me; Richard Edmondson for inviting me to San Francisco Liberation Radio for my first micro radio experience; Chris A., without whose help I would never been able to build KPBJ; Larry, without whose technical expertise, kindness and generosity I would never have been able to stay on the air in San Francisco, let alone set up or run KBLT in Los Angeles; and Pete Tridish for his continuing and tireless work in support of the cause.

In San Francisco, the original cast of DJs: Will, Galen Newman, Danny Clark, Fire, Abby Lewis, Brad Gates, and their friends.

In Los Angeles, Jay Babcock, for putting up with me and being my best friend and biggest supporter while KLBT was on the air; Mark Mauer and Brandon Jay for reaching out to help someone they barely knew and referring their friends to the station; the first wave of DJs to sign up and join me on air: Kerry Murphy, Carolyn Kellogg, Dave Sanford, Chris Carey, Matt Semancik, Miwa Okumura, Fred Kiko, Dale Johnson, Rudy Provencio, Jasmin Segura, Shawn Euzebio, Bradley Temkin, and Sam Wick; the second wave of DJs they brought in: Laura Graven, Bill Smith, Andy Sykora, Maki Tamura, Keith Holland, Courtney Heller, Chris Herrera, Stan Misraje, and Robert Sullivan; and so on: Matt Devine, Steve Gizicki, Wing Ko, Paris Potter, Cake Nuñez, Dana Pilsen, Shawn Kamano, Laurel Stearns, Forest Nelson, AJ Peralta, Julie Hermelin, Jessica Hopper, Franklin Bruno, Brad Laner, John Napier, Jeremy Steckler, Mundo, Mark Kates, Lynnel, Cali, Edith Vache, Tori Horowitz, Pete Relic, David Everett, Kristin Rolla, Amy Blaubaum, Paul Modiano, David Miller, Robert Cappadona, George Barker, Daryl Carlton, Jim Freeman, Alice Chang, Maryann Bowers, Mario Prietto, Doran Meyers, Dean Beckley, Doug Miller, Mo Figuls, Amos Menjiver, Michael Calvert, Eddie Muñoz, Melinda Simon, Mark Fay, Bennett Theissen, Raquel Muñoz-Corbi, Kari French, Paul Greenstein, Doug Glazer, Laura Plansker, Fritz Michaud, Steve K., Gabie Strong, Len Nevarez, Russ & Mok, Bill Mahoney, Jeffrey Plansker, Andrew Blustain, Jim Freek, Paul V., Shondra Bowie, Mike Watt, Keith Morris, Bob Forrest, Howie Klein, Ben Fisher, Marty Sokel, Lida Husik, John Christian, Hassan Jamal, S. A. Griffin, Greg Bishop, Don Bolles, Morgan Higby, Max Levin, David Ponak, Chris Maigatter, Marcel DeJure, Simon Lamb, Eddie Ruscha, Laura Bong, Jacob Cohan, Tony Aronov, Ron Frank, Alyssa Coppelman, Jenn Joos, Rob Miller, Doreen Sanchez, Eric Aguilar, Riley Mohr, Jay Clark, Sandy Rodriguez, Andy Ybarra,

Mark Sovel, Kathy Rivkin, Rich Rubin, Matt Kelly, Mark Farmer, Brian Rosser, Jeremy Bate, and the hundreds of other people who came in as guests or substitutes whose names, unfortunately, I either can't recall or I never knew in the first place.

Chris W., for keeping KBLT technologically afloat, and his friend/consultant, Ed Armstrong; Camille Garcia for all of her amazing art; Stephanie Hernstadt for her photographic expertise; Mark McNeill, for saving us from the FCC the first time they were onto us; the many record labels that supported KPBJ and KBLT by providing free music; the numerous bands who supported us through benefit concerts; the countless other bands who visited the station to spin records and/or play live.

Chris Noxon, Joan Springhetti, Ralph Frammolino, Melinda Simon, and Casey Dolan, for reading the original manuscript and giving me such valuable feedback; Bonnie Nadel, my agent, for seeing the potential in this story and taking it on; Brant Rumble, my editor, for his enthusiasm and keen eye; and Chris Econn, for his love, support, and tireless reading of version after version of this book. Chris, you truly are the best.

CPSIA information can be obtained at www.ICGtesting.com
Printed in the USA
LVOW120248100512

281142LV00002B/49/A